P9-ASJ-004

Apple® Watch For Seniors

by Dwight Spivey

for
dummies®
A Wiley Brand

WITHDRAWN FROM
RAPIDES PARISH LIBRARY

Apple® Watch For Seniors For Dummies®

Published by: John Wiley & Sons, Inc., 111 River Street, Hoboken, NJ 07030-5774, www.wiley.com

Copyright © 2022 by John Wiley & Sons, Inc., Hoboken, New Jersey

Published simultaneously in Canada

No part of this publication may be reproduced, stored in a retrieval system or transmitted in any form or by any means, electronic, mechanical, photocopying, recording, scanning or otherwise, except as permitted under Sections 107 or 108 of the 1976 United States Copyright Act, without the prior written permission of the Publisher. Requests to the Publisher for permission should be addressed to the Permissions Department, John Wiley & Sons, Inc., 111 River Street, Hoboken, NJ 07030, (201) 748-6011, fax (201) 748-6008, or online at http://www.wiley.com/go/permissions.

Trademarks: Wiley, For Dummies, the Dummies Man logo, Dummies.com, Making Everything Easier, and related trade dress are trademarks or registered trademarks of John Wiley & Sons, Inc. and may not be used without written permission. Apple Watch is a registered trademark of Apple, Inc. All other trademarks are the property of their respective owners. John Wiley & Sons, Inc., is not associated with any product or vendor mentioned in this book. *Apple Watch® For Seniors For Dummies®* is an independent publication and has not been authorized, sponsored, or otherwise approved by Apple, Inc.

LIMIT OF LIABILITY/DISCLAIMER OF WARRANTY: WHILE THE PUBLISHER AND AUTHORS HAVE USED THEIR BEST EFFORTS IN PREPARING THIS WORK, THEY MAKE NO REPRESENTATIONS OR WARRANTIES WITH RESPECT TO THE ACCURACY OR COMPLETENESS OF THE CONTENTS OF THIS WORK AND SPECIFICALLY DISCLAIM ALL WARRANTIES, INCLUDING WITHOUT LIMITATION ANY IMPLIED WARRANTIES OF MERCHANTABILITY OR FITNESS FOR A PARTICULAR PURPOSE. NO WARRANTY MAY BE CREATED OR EXTENDED BY SALES REPRESENTATIVES, WRITTEN SALES MATERIALS OR PROMOTIONAL STATEMENTS FOR THIS WORK. THE FACT THAT AN ORGANIZATION, WEBSITE, OR PRODUCT IS REFERRED TO IN THIS WORK AS A CITATION AND/OR POTENTIAL SOURCE OF FURTHER INFORMATION DOES NOT MEAN THAT THE PUBLISHER AND AUTHORS ENDORSE THE INFORMATION OR SERVICES THE ORGANIZATION, WEBSITE, OR PRODUCT MAY PROVIDE OR RECOMMENDATIONS IT MAY MAKE. THIS WORK IS SOLD WITH THE UNDERSTANDING THAT THE PUBLISHER IS NOT ENGAGED IN RENDERING PROFESSIONAL SERVICES. THE ADVICE AND STRATEGIES CONTAINED HEREIN MAY NOT BE SUITABLE FOR YOUR SITUATION. YOU SHOULD CONSULT WITH A SPECIALIST WHERE APPROPRIATE. FURTHER, READERS SHOULD BE AWARE THAT WEBSITES LISTED IN THIS WORK MAY HAVE CHANGED OR DISAPPEARED BETWEEN WHEN THIS WORK WAS WRITTEN AND WHEN IT IS READ. NEITHER THE PUBLISHER NOR AUTHORS SHALL BE LIABLE FOR ANY LOSS OF PROFIT OR ANY OTHER COMMERCIAL DAMAGES, INCLUDING BUT NOT LIMITED TO SPECIAL, INCIDENTAL, CONSEQUENTIAL, OR OTHER DAMAGES.

For general information on our other products and services, please contact our Customer Care Department within the U.S. at 877-762-2974, outside the U.S. at 317-572-3993, or fax 317-572-4002. For technical support, please visit https://hub.wiley.com/community/support/dummies.

Wiley publishes in a variety of print and electronic formats and by print-on-demand. Some material included with standard print versions of this book may not be included in e-books or in print-on-demand. If this book refers to media such as a CD or DVD that is not included in the version you purchased, you may download this material at http://booksupport.wiley.com. For more information about Wiley products, visit www.wiley.com.

Library of Congress Control Number: 2021948931

ISBN 978-1-119-82839-6 (pbk); ISBN 978-1-119-82840-2 (ebk); 978-1-119-82841-9

SKY10030248_111121

Contents at a Glance

Table of Contents

Introduction

Apple Watch is fast becoming a cultural icon in the same vein as other Apple hits, such as iPhone and AirPods. As far as smartwatches are concerned, Apple Watch has no equal and, frankly, not even a close second. More and more folks are adorning their wrists with them every day. Apple Watch is easy to use, works seamlessly with other Apple products, and just plain looks good.

I've used Apple Watch since its first iteration (I can honestly say I got one of the first off the assembly line), and have been a satisfied customer ever since. In this book, I share my expertise with you, to help get you up to speed quickly and discover all the features your Apple Watch has to offer.

About This Book

Like other *For Seniors For Dummies* books I've authored, this one is written for the mature folks among us. People who may be somewhat new to using a smartwatch and want to find out just what these little wonders can do. From turning on and charging your Apple Watch, to customizing watch faces, syncing content with your iPhone, connecting to a Wi-Fi network, holding voice and text conversations, checking email, finding your way around town, and more, *Apple Watch For Seniors For Dummies* tries to cover it all. In writing this book, I made it my mission to consider the types of activities that would interest someone who is 50 years old or older (as I am) and donning an Apple Watch for the first time.

Foolish Assumptions

This book is organized by tasks. Starting right from the beginning, I assume you've never touched, let alone used, an Apple Watch. I also assume you're at least somewhat familiar with an iPhone, since you'll need to have one to use your Apple Watch. (If not — and here comes the shameless plug — a very fine introductory tome called *iPhone For Seniors For Dummies*, written by yours truly, will get you up to speed fast.) Even though you may be tech-savvy to a lesser or greater degree, I try to use nontechnical language throughout the book.

Another assumption I'm making is that you can't wait to find out more about using your Apple Watch to

» Keep track of your workouts

» Stay on top of your current health conditions by monitoring heart rate, blood oxygen levels, sleep tracking, ECG, and more

» Translate into different languages

» Listen to music, audiobooks, and podcasts

» Get directions

» Pay for stuff using Apple Pay

And that's just scratching the surface.

Icons Used in This Book

Icons are the tiny pictures in a page's margin that call your attention to special advice or information. Following are the icons in this book.

TIP

These brief pieces of advice help you to take a skill further or provide alternate ways of doing things.

WARNING

Heads up! This icon flags a task to perform with care — because a mishap would prove difficult or expensive to undo — or a danger you might not be aware of.

REMEMBER

This icon indicates information that's so useful, it's worth keeping in your head, not just on your bookshelf.

TECHNICAL STUFF

Maybe the information here isn't essential, but it's neat to know if you want to impress someone in your inner circle.

Beyond the Book

Even more Apple Watch information is on www.dummies.com. This book's cheat sheet shows you how to extend your Apple Watch's battery life, offers assistance with troubleshooting an unresponsive Apple Watch, and points you in the right direction for Apple support. To get to the cheat sheet, go to www.dummies.com and type Apple Watch For Seniors For Dummies Cheat Sheet in the Search box. This is also where you'll possibly find any significant updates or changes that occur between editions of this book.

Where to Go from Here

I wrote this book in such a way that you can go straight through from beginning to end or skip to a particular chapter to learn a specific topic or immediately work with a certain task. The steps in every task quickly walk you through the process, without bogging you down with a lot of technical jargon.

At the time I wrote this book, all the information contained within was accurate for Apple Watch Series 3, 4, 5, 6, and 7, Apple Watch SE, and version 8 of watchOS (the operating system used by the Apple Watch). It's likely that Apple will introduce new Apple Watch series and versions of watchOS between editions of this book. If you've bought a new Apple Watch and found that its software, hardware, or user interface, or other software on your computer looks or acts a little different than what's in the text, check out the Apple Watch website at www.apple.com/watch. You'll most likely find updates there on the latest releases and information for Apple Watch.

1
Getting to Know Your Apple Watch

IN THIS PART . . .

Meeting your new Apple Watch

Pairing and updating your Apple Watch

Customizing settings and discovering installed apps

Making Apple Watch accessible

IN THIS CHAPTER

» **Discover what's new in Apple Watch models and watchOS 8**

» **Choose the best Apple Watch for you**

» **Understand what else you need**

» **Find out where to buy an Apple Watch**

» **Explore what's in the box**

» **Take a look at the device**

» **Get help for your Apple Watch**

Chapter **1**

Buying Your Apple Watch

No doubt you've seen little glass-covered watches with the (usually) colorful bands on the wrists of the young, the old, and the in-betweens. You'll often catch the wearers stealing a glimpse at their wrist multiple times over a period of time, but they're not always checking the time. They're checking text messages. Or being informed that they've reached their activity goals for the day. Or perhaps they're being reminded of a pending appointment. Maybe they're viewing a photo of their grandchildren or even holding a phone conversation with them. Or maybe they're translating something they read in a language they don't understand. Or — and how's this for cool — they're administering an electrocardiogram on themselves. Yes, with an Apple Watch, you can do those things and more, right from your wrist. You, dear reader, are about to be immersed in the world of Apple Watch, and I'm grateful to be your guide.

In this chapter, you learn about the various models of Apple Watch, as well as where to buy one. You also explore what's inside the Apple Watch box, find out what the buttons on the side are for, and discover where to get help if you should ever need it — beyond the little book in your hands, that is.

Discover the Newest Apple Watch Models and watchOS 8

Apple Watch gets its functionality from the combination of its hardware and its software operating system (called *watchOS*, which is short for *Apple Watch operating system*). The most current version of the operating system is watchOS 8. It's helpful to understand which new features the latest models and watchOS version offer the Apple Watch wearer (all of which are covered in much more detail throughout this book).

The newest Apple Watch to grace the lineup is Apple Watch Series 7 (shown in **Figure 1-1**). Although other smartwatches are on the market, this one is the smartwatch equivalent of William James Sidis. (I encourage you to look him up if you haven't heard of him.) Others may work with Android devices and some have more of a fitness bent, but none come close to Apple's offering. Sorry, but this is a book about Apple Watch written by an Apple fan for other Apple fans, so there will be no tomfoolery with further discussions of "competitors." I digress. . .

Here are some of the key features of Apple Watch Series 7:

» **Faster charging:** The Series 7 has a new charging architecture and updated magnetic charger that allow for up to 33 percent faster charging than the Series 6 models. If that doesn't sound like a big deal, just wait until you are in a hurry to leave the house but discover that you forgot to charge your Watch during the night (been there, done that). That 33 percent faster rate will be a big plus in that situation.

Image courtesy of Apple, Inc.

FIGURE 1-1

» **Reengineered always-on Retina display:** The display on the Series 7 is the largest display for an Apple Watch model to date, offering more screen area and brightness than the Series 6. As a result, the display is much easier to see than previous models, especially when sunlight is glaring off the screen. Also, the extra 20 percent of screen space over the Series 6 makes apps easier to use. The "always-on" part means you can now access features of your Apple Watch without having to wake it from sleep, as you did with previous models.

» **Exceptional durability:** While Apple has never given us an Apple Watch that would break at the slightest touch, the Watches have progressed in durability over the years. Now, with the Series 7, they're getting closer to indestructability than ever before. (Please don't test that statement on yours, though.) The crystal on the Series 7 display maintains pristine clarity even though it's 50 percent thicker than the Series 6 display, which of course makes it more resistant to cracks.

TIP

You might consider acquiring AppleCare+, which is Apple's extended warranty, currently priced at $49 for Apple Watch SE, $79 for Apple Watch Series 7, or $149 for the far more expensive Apple Watch Hermès and Apple Watch Edition (both are essentially extremely souped-up models of Apple Watch Series 7). AppleCare+ extends the warranties of the respective models by 1 year, and includes up to two incidents of accidental damage (at an extra cost of between $69 to $79, depending on the model). The accidental damage coverage could more than cover the cost of repairing your Apple Watch without it. Visit www.apple.com/support/products/watch/ to learn more.

This book is based on version 8 of watchOS, which is supported for Apple Watch Series 3, Apple Watch Series 4, Apple Watch Series 5, Apple Watch SE, Apple Watch Series 6, and Apple Watch Series 7. This update to the operating system adds many features, including (but most certainly not limited to) the following:

» **New customizable and shareable watch faces:** The watch face is your gateway into your Apple Watch, and it's what you see first when you raise your wrist. You'll want your watch face to work the way you want — and to look snazzy to boot. watchOS 8 provides new watch faces that you can also customize and share with other Apple Watch fans. These new faces are also optimized to utilize the larger screen area of the Series 7 models. More on this topic in Chapter 5.

» **Multiple complications:** In previous watchOS versions, you were limited to a single complication per app. Now, apps are able to offer multiple complications at once, allowing you even greater customization of your watch face. What in the world is a *complication*? See the nearby sidebar.

» **Fall detection enhancements:** Your Apple Watch is so advanced that it can detect when you take a hard fall, and even contact emergency services if needed. watchOS 8 includes updates and enhancements for fall detection that make this functionality more accurate and work better for specific types of falls (such as falling from a bicycle).

WHAT IN THE WORLD IS A COMPLICATION?

Have you ever noticed the tiny little windows and miniature hands and faces that litter the larger face of some analog watches? These little doo-dads are called *complications* (as shown in the corners of the watch face in the figure), and they allow your watch to provide you much more information than just the time (such as the date or moon phases, for examples). watchOS has always provided complications for your Apple Watch faces. Those complications may be part of watchOS or they may come with other third-party apps that you install on your Apple Watch.

Image courtesy of Apple, Inc.

» **Mindfulness app:** The Mindfulness app, which helps you focus and center yourself throughout your day, replaces and incorporates the breathe mode of previous watchOS versions. The breathe functionality guides you through breathing exercises, but now a second function, Reflect, also helps you exercise your mind and soul.

TIP

Not all features of watchOS 8 work with every Apple Watch model. If you have an older Apple Watch, you may not be able to use certain watchOS 8 features due to hardware limitations (such as the new watch faces designed for the larger screen areas on the Series 7 models).

These are but some of the improvements made to the latest version of watchOS. Please consider visiting `www.apple.com/watchos/watchos-8` to discover more.

Choose the Right Apple Watch for You

There's an Apple Watch out there that's just right for you, trust me. This gadget is the most customizable Apple's ever offered, and there's a style to suit every taste.

Apple sells several versions of Apple Watch at various price and consumer focus points. Apple Watch Series 7 is the newest model, so that will be the primary focus of this section, but I will make some comparisons between it, the Apple Watch SE, and the Apple Watch Series 3 (which are two other models Apple currently sells).

Apple Watch Series 7 models come in 41mm and 45mm case sizes, SE models come in 40mm and 44mm, and Series 3 comes in 38mm and 42mm.

TIP

You might think "the larger the better" applies here when it comes to tapping on the screen of your Apple Watch, and that may be true if you have larger wrists and fingers. However, small wrists have necessitated my use of the 40mm and 38mm cases for years, and I've had no trouble at all.

When selecting your Apple Watch, you'll need to select not only a size for your case but a wealth of other choices:

>> **Case material:** Apple Watch Series 7 comes in aluminum, stainless steel, and titanium.

>> **Case color:** Apple Watch Series 7 offers more case colors than ever (depending on the material you select):

- Midnight (aluminum)

- Starlight (aluminum)

- Green (aluminum)

APPLE WATCH COLLECTIONS

Apple also offers Apple Watch in collections, including Apple Watch Nike and Apple Watch Hermès, both shown in the figure. Apple Watch Nike (`www.apple.com/apple-watch-nike/`) allows you to select from several unique Nike-designed bands and includes special apps focused on Nike fitness activities. Apple Watch Hermès (`www.apple.com/apple-watch-hermes/`) is the upper echelon of Apple Watch models. It comes with a unique watch face and you can choose from an extensive array of gorgeous bands and clasps ranging from top-of-the-line fabrics to luxurious leathers.

Apple Watch Nike

Apple Watch Hermès

Image courtesy of Apple, Inc.

- Blue (aluminum)
- Product RED (aluminum)
- Silver (stainless steel)
- Gold (stainless steel)

- Graphite (stainless steel)
- Space Black (titanium)
- Titanium (titanium)

» **Band:** Selecting a band might be your most time-consuming task, as there is no shortage of colors (you name it!), styles (loops, bands, wraps, and more) and material types (rubber, leather, metals, fabrics, and so on) to choose from.

» **Cellular or non-cellular:** Every Apple Watch has GPS, but you can also opt for the GPS + cellular model, which allows you to use it with or without your iPhone for calls, texts, and more activities that require a data connection. Visit www.apple.com/watch/cellular/ to find a full list of cellular carriers for the Apple Watch model you'd like to use.

Other differences between Apple Watch models come primarily from the current operating system, watchOS 8. Newer models, such as Apple Watch Series 7, support some tasks and older models do not.

Table 1-1 gives you a quick comparison of Apple Watch Series 7, Apple Watch SE, and Apple Watch Series 3 (models currently sold by Apple). All costs were current when this book was written. (Some carriers may introduce non-contract terms.)

TABLE 1-1   **Apple Watch Model Comparison**

Model	Materials	Sizes	Cost	Carriers
Series 7	Aluminum, stainless steel, titanium	41mm and 45mm	from $399	AT&T, Verizon, T-Mobile, and others (check out www.apple.com/watch/cellular/#table-series-7 for the full list)
SE	Aluminum only	40mm and 44mm	from $279	AT&T, Verizon, T-Mobile, and others (see www.apple.com/watch/cellular/ – table–se for an exhaustive list)
Series 3	Aluminum only	38mm and 42mm	from $199	Cellular not available

TIP For a side-by-side comparison of all three models Apple currently sells, check out www.apple.com/watch/compare/.

Understand What You Need to Use Your Apple Watch

Before taking the plunge with a new Apple Watch, you should know what you'll need to take full advantage of its wares.

Right off the bat, you'll need an iPhone (even GPS + cellular versions require you have one) — no ifs, ands, or buts. The iPhone is necessary for updating your Apple Watch's operating system (watchOS), installing apps, setting up optional services such as Apple's Fitness+, and more. You simply must have an iPhone, along with its Apple Watch app, if you want to set up and get going with your Apple Watch. If you have a GPS + cellular model, you still need the iPhone to get started before you can begin to use the Apple Watch without it. The three Apple Watch models Apple currently sells require an iPhone 6 or later running at least iOS 15.

TIP Family Setup is an option in watchOS that allows you to use a single iPhone to set up multiple Apple Watches. That way, everyone in the family won't need their own individual iPhone. For more info on this great feature, visit https://support.apple.com/en-us/HT211768.

If you want to be able to use your Apple Watch GPS + cellular model without an iPhone (aside from the requirement to have one for setting up, as mentioned in the preceding paragraph), you'll need to add your Apple Watch to the data plan account you have with your cellular provider. The data plan allows you to exchange information over the internet (such as emails and text messages) and download content (such as music). Try to verify the strength of coverage in your area, as well as how much data your plan provides each month, before you sign up.

You should already have a free iCloud account (since you have or will get an iPhone), Apple's online storage and syncing service, to store and share content online among your Apple devices, keep track of your devices, and more. For example, you can set up iCloud in such a way that you can find Apple devices you've lost, which is handy if you happen to misplace your Apple Watch.

Find Out Where to Buy Your Apple Watch

You can buy an Apple Watch from just about anywhere, including local retailers and online. You can find them at a brick-and-mortar or online Apple Store, from mobile phone providers, such as AT&T, Sprint, T-Mobile, and Verizon, and at major retailers, such as Best Buy and Walmart, but the choice of models and bands may be limited to on-hand supplies. You can also find Apple Watch at several online retailers (such as Amazon.com and Newegg.com) and through smaller, local stores and shops.

However, if you want to explore the full range of Apple Watch models and options, visit Apple Watch Studio found at `www.apple.com/shop/studio/apple-watch`. If you don't need your Apple Watch this very instant, Apple Watch Studio is your best bet to find the Apple Watch that meets your every wish in wearable wrist technology. Click the blue Get Started button and customize to your heart's content.

TIP

Apple also offers an easy way to find your nearest Apple Watch retailer. Visit `https://locate.apple.com/`, click Sales, enter your address or ZIP code, select Apple Watch from the Products menu, and click Go to see a list of local establishments.

See What's in the Box

When you fork over your hard-earned money for your Apple Watch, you'll be left holding a minimalist-inspired long, slender box. Box details and colors may vary, depending on the series you purchase.

Guess what you'll find when you take off the shrink wrap and open the box? More boxes, of course!

One box contains:

TIP

- » **Your Apple Watch:** Save all the packaging until you're certain you won't return the Apple Watch. Apple's standard return period is 14 days, but only for products you purchase directly from them. If you purchase your Apple Watch from a third party, check with them for their return policies.

- » **Documentation (as with other Apple products, I use the term loosely):** The documentation typically includes a small pamphlet, a sheet of Apple logo stickers, and a few more bits of information.

- » **USB-A charging cable:** Use this cable (see **Figure 1-2**) to connect the Apple Watch to your computer or USB power adapter (not included) for charging.

The second box contains your watch bands. Be on the lookout for the little green dot on one end of the box, which indicates the tab you pull to open it.

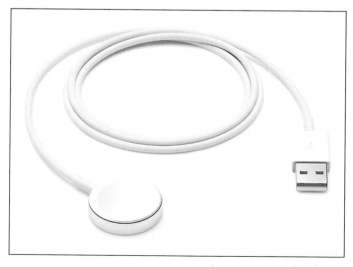

Image courtesy of Apple, Inc.

FIGURE 1-2

TIP

You can search for Apple Watch accessories online. You'll find covers and cases, bands, headphones, stands, screen guards, and much more. Apple's own website should be your first stop for such accessories (www.apple.com/shop/watch/accessories), but don't limit yourself to only that site. There's a seemingly endless array of worthy retailers vying for your business.

Take a First Look at Apple Watch

In this section, I give you a bit more information about the buttons and other physical features of the newest Apple Watch models. **Figures 1-3** shows you the location of these items on the Apple Watch Series 6, Apple Watch SE, and Apple Watch Series 3, respectively.

Here's the rundown on what the various hardware features for the given Apple Watch models and what they do:

» **Side button:** Turn your Apple Watch on or off, show or hide the dock, use Apple Pay, or use the emergency SOS feature.

» **Digital crown:** Launch Siri, double-click to return to the previous app, scroll up and down lists, zoom in and out, and more.

» **Display:** See where the cool stuff appears in the window on your Apple Watch.

» **Electrical heart sensor (Series 6 only):** Measure electrical signals in your blood, allowing your watch to perform an ECG and provide accurate heart health diagnostics.

» **Optical heart sensor:** Measure how quickly your heart beats and your blood flows. Although not as accurate as an electrical heart sensor for some measurements, the optical heart sensor is handy during workouts.

» **Blood oxygen sensor (Series 6 only):** Measure your blood's oxygen saturation. Using clusters of red, green, and infrared lights, the sensor (shown in **Figure 1-4**) detects the color of the blood flowing through your veins to determine how much oxygen it contains.

Apple Watch Series 7

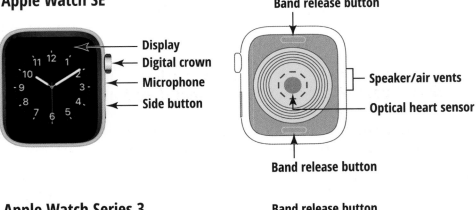

Apple Watch SE

Apple Watch Series 3

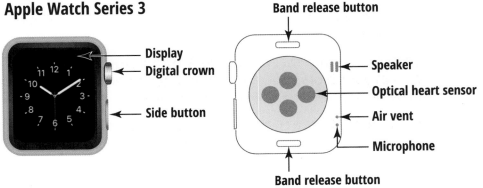

FIGURE 1-3

AREN'T THERE THREE NUMBERS BETWEEN 3 AND 7?

By now, with all the mention of the Apple Watch Series 3 and Apple Watch Series 7, you may be wondering whether Apple simply made an illogical jump in numbering or whether I've determined to hide the fact that the numbers 4, 5, and 6 exist. I assure you, it's neither. Rather, Apple offers the Series 3 as a capable albeit older (and much less expensive) alternative, the SE as a newer but less expensive offering, and the Series 7 as their latest and greatest standard-bearer. Series 4, Series 5, and Series 6 are great smart watches (and can still be found for purchase from many third-party resellers), but the SE and Series 7 cover the tech bases for both models.

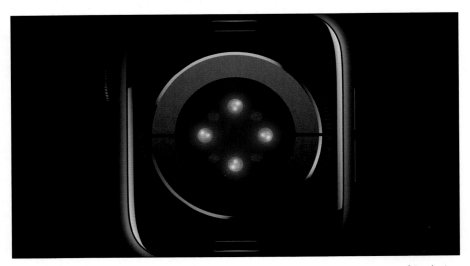

Image courtesy of Apple, Inc.

FIGURE 1-4

» **Speaker/air vents:** Hear audio output and equalize pressure to assist the altimeter in providing more accurate altitude measurements. (The speaker and air vents are two separate components on the Series 3.)

- » **Band release buttons:** Press to release the bands connected to your Apple Watch, and then slide the bands out to remove them.

- » **Microphone:** Speak into your Apple Watch to deliver commands or content. This feature allows you to do such things as make phone calls, dictate text messages, and work with other apps that accept audio input, such as the Siri built-in assistant.

Get Further Help for Your Apple Watch

The book you hold in your hands is *the* one (in my humble opinion) you'll need to get started with your Apple Watch and to understand how to use it for your purposes. However, I can't pretend that it covers everything that's possible to know or understand about your *Star Trek*-ish device. For example, you may be an engineer who wants to know how apps are developed for Apple Watch — I don't dabble in software development in *Apple Watch For Seniors For Dummies*. Or perhaps you want to get in touch with Apple if the unfortunate happens and you need help with troublesome hardware that won't work as it should — I tell you where to find such help in this section.

Here are several helpful resources every Apple Watch owner should be aware of:

- » Apple's official Apple Watch support website can be found at `https://support.apple.com/watch`.

- » Apple's official support contact website is at `https://support.apple.com/contact`.

- » The phone number for reaching Apple support in the United States is 1-800-275-2273. For Canada, call 1-800-263-3394. To find numbers for other regions, visit `https://support.apple.com/en-us/HT201232`.

- » If you prefer to contact Apple for shopping or support with an ASL interpreter, check out `https://signtime.apple/applecare/us-EN/asl`.

» Does your Apple Watch need servicing or do you simply want to check your warranty status? Visit Apple's Apple Watch service and repair website at `https://support.apple.com/watch/repair/service`.

» Do you need help finding the serial number or IMEI (International Mobile Equipment Identity) for your Apple Watch? This article will help: `https://support.apple.com/en-us/HT204520`.

» Please do take the time to review the following website for Apple Watch safety information: `https://support.apple.com/guide/watch/important-safety-information-apdcf2ff54e9/7.0/watchos/7.0`. You'll find a good deal of info on how to properly use and store your investment both for its protection and, more importantly by far, for your own.

» And for the aforementioned engineers in the crowd who can't resist the urge to dig deeper into the how's of Apple Watch, here you go: `https://developer.apple.com/watchos/`. Have fun!

ii none

» **Learn how to navigate the Watch**

» **Charge your Apple Watch**

» **Connect to Wi-Fi and Bluetooth**

» **Customize with the Apple Watch app**

» **Wake and turn off Apple Watch**

Chapter **2**

Setting Up

Friends, it's time to get down to business with that lovely timepiece you've purchased from the good folks at Apple. They've worked hard to give you a device worth using, and you've put in your own work researching and then shelling out your hard-earned cash for this little marvel.

In this chapter, you find out the basics of using your Apple Watch. You learn how to set up your Watch, pair it with your iPhone, navigate its features, customize its settings, and generally make this work of art on your wrist work, look, and act just like you want it to.

Let's get started!

Set Up a New Apple Watch

Your Apple Watch is a small wonder, but in spite of all the incredible technology crammed into that tiny space, it still needs a bit of help from your iPhone to get started. Be sure to keep your iPhone handy if you're following along with the steps in this chapter.

REMEMBER

You need an iPhone 6s or newer running iOS 15 or newer to pair with your Apple Watch.

Put on your Apple Watch

First things first: It's time to put your Apple Watch on your wrist. Which wrist you use is up to you.

This part sounds like a no-brainer, right? Well, yes and no. Yes, because using the strap on the Watch is easy to figure out. No, because you should make sure that the Watch isn't too loose; if it is, the sensors on it may not work as intended and the accuracy of measurements such as blood oxygen will be dramatically reduced.

Make sure that your Apple Watch is snug enough to touch your skin, but not so snug that your hand begins to turn shades of red and purple.

Apple goes into greater detail about how you should properly fit and wear your Watch, and lists the materials in their various models for those who are sensitive to certain materials. Visit `https://support.apple.com/en-us/HT204665` for more information.

Turn on and pair your Watch

Once you've got your Watch on your wrist, it's time to start it up!

1. Press and hold down the side button on your Apple Watch until the Apple logo pops up on the display.

2. Hold your iPhone close to the Apple Watch so the two can become better acquainted.

After the devices find one another, the pairing process begins.

TIP

Make sure Bluetooth is enabled on your iPhone in Settings⇨Bluetooth before beginning the pairing process. This is the beginning of a beautiful relationship; you don't want to get off to a bumpy start.

3. When the pairing screen pops up on your iPhone, tap the Continue button.

TIP

If you don't see the pairing screen after a minute or two, open the Apple Watch app on your iPhone and tap the Pair New Watch button.

4. When prompted, move your iPhone so that your Apple Watch appears in the viewfinder window on your iPhone (see **Figure 2-1**).

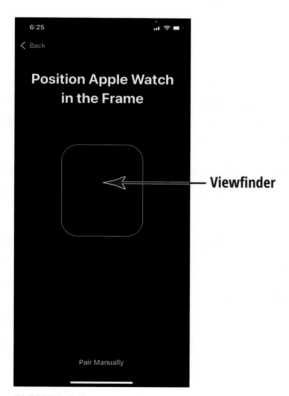

FIGURE 2-1

5. Tap the Set Up for Myself button when you see it, and then follow the onscreen instructions to complete the setup and pairing process.

DO YOU HAVE MORE THAN ONE APPLE WATCH?

You can pair multiple Apple Watches with a single iPhone. Go to the Apple Watch app on your iPhone to add a Watch, switch between Watches, customize your Watch, and more. For more information, head over to this page on Apple's Support site: https://support.apple.com/en-us/HT205792.

TIP

What to do if your Apple Watch and iPhone just won't pair up? Visit Apple's Support site at https://support.apple.com/en-us/HT205025 for help with that dilemma.

Navigating your Apple Watch

Once you Watch is paired and ready to go, you'll want to know how to move around this new miracle. (Okay, *miracle* may be a bit strong, but how else to explain how Apple crammed all this great stuff into such a tiny box?)

You need to know only a few basic maneuvers and gestures to peruse your Apple Watch's wares:

» Use your finger to tap icons or buttons on the display. This simple action opens apps, makes option selections, sends messages, and more. Sometimes, you may need to tap with more than one finger, or even tap more than once, but those gestures are for more advanced tasks described later in the book.

» Use your finger to swipe up, down, left, and right on the display. This movement scrolls through options, zips between multiple screens in an app, and more.

» Use your finger to drag an item on the display. This comes in handy when panning around an image and finding an app on the Home screen (see **Figure 2-2**).

FIGURE 2-2

» Rotate the digital crown. You use this move when you want to quickly scroll up and down through options or when changing apps.

» Press the digital crown to return to the Home screen from anywhere. Press it again while on the Home screen to open the Clock app.

» Press the side button once to open the dock, as shown in **Figure 2-3**, and view apps you've used recently. (For more on the dock, see Chapter 3.)

» Double-click the side button when using Apple Pay. (See Chapter 9 for much more info on Apple Pay.)

There are other combinations and variations of these movements and gestures for specific tasks, but this more than covers the bases for now. More on navigation when a task warrants it.

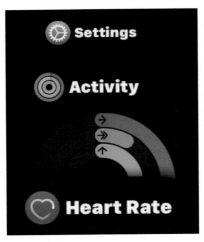

FIGURE 2-3

Keep Your Apple Watch Running

Apple Watch batteries can last quite a while, but they're not infinitely charged, I'm sad to say. That's why a charger is super handy.

The Apple Watch charger, shown in **Figure 2-4**, magnetically connects to the back of your Apple Watch and delivers charge to the battery without plugging into the Watch.

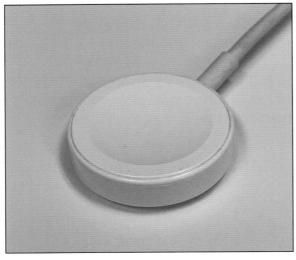

FIGURE 2-4

Charge your Apple Watch

Here's how to charge your Apple Watch:

1. Connect the charger to a power supply.

   The other end of the charger (the one not seen in Figure 2-4) is a standard USB-A connector. You can use a USB-A power adapter (which you would, of course, plug into a power outlet) to connect the charger, or you can plug the charger cable into a USB-A port on your computer.

2. Attach your Apple Watch to the charger end, shown in **Figure 2-5**.

   When charging begins, you'll see a charging icon on the Apple Watch display (see **Figure 2-6**).

FIGURE 2-5

FIGURE 2-6

3. As the Watch charges, you can see its progress by tapping the display.

   A circle showing the current amount of charge appears. Once the charge is at 100%, you'll see that the progress circle is complete, like the one shown in **Figure 2-7**.

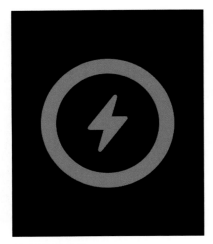

FIGURE 2-7

Keep tabs on the battery power

As you go through the day (or night), you may want to check out just how much charge is remaining for your battery. You can easily do so in a couple of ways: using Control Center or using a watch face complication.

To check your charge in Control Center:

1. Swipe up from the very bottom of the Apple Watch display to open Control Center.

2. The battery percentage button in Control Center, shown in **Figure 2-8**, displays the amount of charge remaining.

If the charge is low and you want to preserve it, tap the battery percentage button, slide the Power Reserve button (shown in **Figure 2-9**) from left to right, and then tap the Proceed button to enter power reserve mode. This mode turns off all features except time keeping. To exit power reserve mode, simply hold down the side button.

If the watch face you're using has a battery complication, like the one in **Figure 2-10**, you can use it to keep track of the remaining charge.

You can tap the complication to open the same window shown in Figure 2-9, as well.

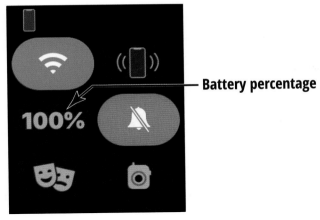

Battery percentage

FIGURE 2-8

FIGURE 2-9

Battery complication

FIGURE 2-10

TIP

Chapter 5 has much more to say about watch faces and complications.

What to do when your battery is low on power

You can ask your Apple Watch to do quite a bit less than it's capable of if you'd like to preserve its remaining battery charge by using its power reserve mode.

Power reserve mode allows Apple Watch only to display the time. Yep, it turns it into just a plain old (but darned good–looking) digital watch. No other apps are running or can be used.

To place your Watch in power reserve mode:

1. Swipe up from the bottom of the Apple Watch display to open Control Center.

2. Tap the battery percentage button in Control Center.

3. Drag to slide the Power Reserve button, shown in **Figure 2-11**, from left to right.

4. Tap the Proceed button at the bottom of the display (see **Figure 2-12**), and your Apple Watch will take a well-deserved break to save the power it has remaining.

FIGURE 2-11

FIGURE 2-12

5. To exit power reserve mode, press and hold down on the side button until the Apple logo appears on the display.

   You're basically restarting your Apple Watch with this maneuver.

REMEMBER

Apple Watch must have at least 10 percent charge to exit power reserve mode.

WARNING

Don't get complacent using power reserve mode. Even with the mode enabled, eventually your Apple Watch will run out of charge.

When your Apple Watch's battery power reaches a critically low level (10 percent), you'll receive a notification on the display and be offered the chance to enter power reserve mode. If you continue to run the Apple Watch down to its last tiny bit of charge, it will simply put itself into power reserve mode.

Optimize battery charging

One day, a very long time from now (typically), your Apple Watch battery will begin to wear out from use. Apple has cleverly devised a way to ward off that day by optimizing how your Watch battery is charged. Basically, your Apple Watch can learn your charging patterns and make your battery last longer.

To enable this feature:

1. Press the digital crown to access the Apple Watch Home screen.

2. Tap the Settings app (shown in **Figure 2-13**) to open it.

3. Rotate the digital crown or swipe the display to find Battery, and then tap it.

4. Tap Battery Health.

5. Rotate the digital crown or swipe the display until you see the Optimized Battery Charging switch, shown in **Figure 2-14**. If the switch is gray, tap it to turn on the feature.

6. Press the digital crown once to exit Settings and press it again to return to the clock.

TIP

For much more detailed information than I can provide here, please head over to Apple's Support site and check out `https://support.apple.com/en-us/HT210551`. It's worth your time to understand how your Watch's battery works so that you can get the most out of your investment.

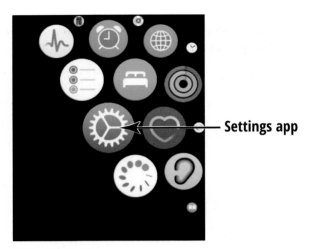

— Settings app

FIGURE 2-13

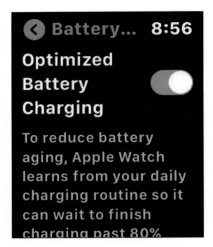

FIGURE 2-14

Connecting Apple Watch to Wi-Fi

As a modern device, your Apple Watch loves to connect to a Wi-Fi connection so it can hop on the internet and talk to other devices connected to the Wi-Fi network. Wi-Fi is also the main way your Apple Watch will stay connected to the world around it if your iPhone isn't in the near vicinity (or is turned off).

To connect Apple Watch to Wi-Fi:

1. Press the digital crown to access the Apple Watch Home screen.

2. Tap the Settings app (refer to Figure 2-13) to open it.

3. Rotate the digital crown or swipe the display to find the Wi-Fi button, and then tap it.

4. If necessary, toggle the Wi-Fi switch on (green), as shown in **Figure 2-15**.

5. Rotate the digital crown or swipe the display, if necessary, until you see the name of the Wi-Fi network you want to join, and tap it.

   If the network requires a password, you'll see a screen like the one in **Figure 2-16**.

FIGURE 2-15

FIGURE 2-16

6. If you need to enter a password, use either the scribble tool or your iPhone's keyboard:

   - *Scribble:* Tap the scribble button (a finger pointing to a letter), and drag your finger on the display in the shape of the first letter in the password. Each time you do so, if Apple Watch recognizes the character you've scribbled, it will add it to the password field at the top of the display. Rotate the digital crown if you need to change a character to uppercase or lowercase. After you've scribbled the entire password, tap the blue Join button in the upper right.

   - *Keyboard:* Tap the button that looks like a keyboard and you'll see the prompt in **Figure 2-17**. When you see the message *Apple Watch Keyboard Input* on your iPhone screen, tap to open a keyboard on its screen. Enter the Wi-Fi network's password and tap the blue Join button to join it.

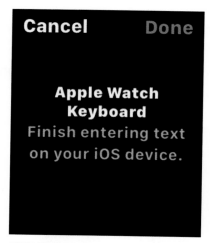

FIGURE 2-17

Connect Bluetooth Devices to Apple Watch

Bluetooth is not the latest scary thing your dentist wants to talk to you about. Rather, it's a wireless communication technology that lets you connect cool things like earbuds, headphones, and speakers to your Apple Watch (and other devices that support Bluetooth, such as your iPhone), with nary a single wire.

REMEMBER

Bluetooth must be enabled on both your Apple Watch and your iPhone for them to maintain their connection.

Pair a Bluetooth device

To connect Apple Watch and Bluetooth devices:

1. Press the digital crown to access the Apple Watch Home screen.

2. Tap the Settings app (refer to Figure 2-13) to open it.

3. Rotate the digital crown or swipe the display to find Bluetooth, and then tap it.

4. If necessary, toggle the Bluetooth switch on (green), as shown in **Figure 2-18**.

   When Bluetooth is on and Bluetooth devices are in range, they'll appear in a list, as shown in **Figure 2-19**.

5. Tap a device to connect to it.

FIGURE 2-18

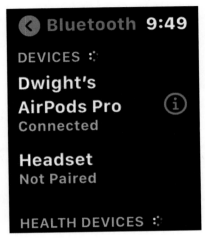

FIGURE 2-19

Select an audio output source

A speaker is built into your Apple Watch, but if you want to change your audio output to another device (such as a Bluetooth device), you can do so easily.

To change audio output sources:

1. Swipe up from the bottom of the Apple Watch display to open Control Center.

2. Tap the audio output button in Control Center (labeled in **Figure 2-20**).

3. Tap the audio output source you want to use, an example of which can be seen in **Figure 2-21**.

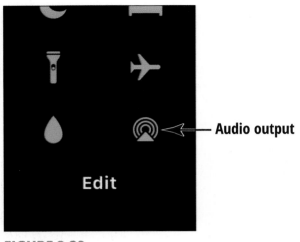

Audio output

FIGURE 2-20

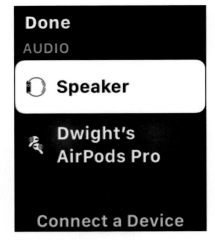

FIGURE 2-21

Meet the Apple Watch App

You've already used the Apple Watch app on your iPhone to pair it with your Watch. Besides this handy little trick, the Apple Watch app can customize your Watch's settings, create watch faces, update software, and so much more. While a lot can be done from the Watch itself, let's be honest: It's much easier to do things on the iPhone's larger screen than it is on the itty-bitty display of your Apple Watch.

The Apple Watch app, shown in **Figure 2-22**, can be found amongst the other apps on your iPhone's Home screen. Tap to open this treasure chest of an app.

The app opens and displays the name of the Watch it's currently using at the top of the screen (see **Figure 2-23**). If not, make sure the My Watch tab is selected at the bottom of the screen.

You may feel a bit overwhelmed when you start to browse around the Apple Watch app because there's a lot to take in. If you can do a task with your Apple Watch, you'll find a setting for it in the Apple Watch app too. Don't be dismayed; I'll be happy to show you around.

Apple Watch app

FIGURE 2-22

FIGURE 2-23

REMEMBER

I use the Apple Watch app throughout this book to show you how to perform various tasks, so the following is just an overview. The in-depth stuff comes later, trust me.

The My Watch tab

While you're in the My Watch tab, you might as well stay put and look around.

Change or add Watches

If you have multiple Apple Watches connected to your iPhone, you can select the one you want to work with by tapping All Watches in the upper-left corner of the screen (refer to Figure 2-23). The screen shown in **Figure 2-24** appears.

From here, you can tap the name of the Watch you want to use or tap the Add Watch button to add another Watch to this iPhone.

Tap Done to go back to the main My Watch tab page (refer to Figure 2-23).

Manage your watch faces

In the My Faces section of the My Watch tab, you can browse the watch faces on your Apple Watch. Swipe left or right to view the various faces, or tap Edit (refer to Figure 2-23) in the My Faces section to manage those faces, as shown in **Figure 2-25**.

FIGURE 2-24   **FIGURE 2-25**

To manage faces:

» To change a face's position in the list, drag the handle (labeled in Figure 2-25).

>> To delete a watch face from the list and your Apple Watch, tap the red circle to the left of the face, and then tap the red Remove button that appears on the right (refer to Figure 2-25).

Tap Done in the upper right when you've finished managing your watch faces.

Customize device settings

Tap any of the device settings listed to enable or disable features, customize how certain tasks work, and much more. These settings are

>> Notifications

>> App View

>> Dock

>> General

>> Display & Brightness

>> Accessibility

>> Siri

>> Sounds & Haptics

>> Passcode

>> Emergency SOS

>> Privacy

All of these settings are discussed in more detail in various chapters throughout this book.

Customize app settings

Tap any of the app settings listed to customize how the apps you have installed on your Apple Watch operate. Any app that comes with Apple Watch, as well as those you've added, will appear in the My Watch tab. Simply tap one to see its options and features, as I did with the Sleep app in **Figure 2-26**.

FIGURE 2-26

Tap Back in the upper left of the screen when you're ready to move to another app or area of the Apple Watch app.

TIP

Chapter 3 goes into much more detail about apps on Apple Watch, including how to add your own.

Add more apps

The Available Apps section (the last section in the My Watch tab) enables you to install on your Apple Watch apps you already have on your iPhone, assuming the app developer has created an Apple Watch version. Simply tap the Install button to add the app to your Apple Watch.

The Face Gallery tab

Tap the Face Gallery tab at the bottom of the Apple Watch app to see a cornucopia of watch faces you can add to your Apple Watch, as shown in **Figure 2-27**.

Some of these faces aren't installed on your Apple Watch, while others are preconfigured versions of those faces. The preconfigured faces are already customized versions of an existing face, which can save you an awful lot of time if you happen to find a version you like.

Tap one of the customized faces, and then tap the Add button under its name (see **Figure 2-28**) to add it to your repertoire.

There's more in Chapter 3 on this topic.

FIGURE 2-27

FIGURE 2-28

The Discover tab

The Discover tab at the bottom of the Apple Watch app, shown in **Figure 2-29**, is like a reference for your Apple Watch. You'll find topics to help get you started and to discover interesting things you can do with your device.

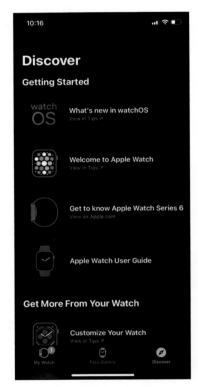

FIGURE 2-29

Tap a topic and be whisked away to a presentation, video, or website that discusses it further.

Wake Up and Turn Off Apple Watch

To wrap up this chapter, let's find out how to wake your Apple Watch from its slumber and how to turn it off (if you ever need to).

Wake up, Apple Watch!

Waking your Apple Watch is easy, since it never really falls into a deep sleep.

To wake Apple Watch, do one of the following:

» Raise your wrist.

» Tap the display.

» Press the digital crown.

» Rotate the digital crown upward.

» Press the side button.

Apple Watch will go back to sleep again when you lower your wrist.

TIP

Apple Watch Series 5, Series 6, and Series 7 include the Always On feature, which allows your display to keep displaying the time even when your wrist is down. If you have one of these models, don't be surprised if the display is on when the watch is sleeping.

Turn it off

To turn off your Apple Watch:

1. Press and hold down the side button until the sliders in **Figure 2-30** appear.

2. Drag the Power Off slider all the way from left to right, and your Apple Watch will power down.

FIGURE 2-30

IN THIS CHAPTER

» **Update the OS**

» **See which apps come with watchOS**

» **Organize and install apps**

» **Explore the dock and Control Center**

» **Discover haptics**

Chapter **3**

Getting Going

The basics have been covered, so now it's time to dive a little deeper into the warm and inviting waters of the pool of Apple Watch knowledge!

Apple Watch is able to do what it does because of the seamless harmony between the software it runs and the hardware that runs it. Since Apple is the only architect of the device, it simply works and does so quite well. What separates the Apple Watch from just another digital timepiece is the veritable cornucopia of apps that connect your Apple Watch to your world (and to your iPhone, to some degree). Apple also ensures that third-party apps compatible with Apple Watch and watchOS meet stringent standards, which is why nearly all apps, Apple or non-Apple, function so flawlessly (or darned near it, anyway).

In this chapter, you find out how to keep your Apple Watch's software and apps up-to-date, discover the apps that are available to you, learn how to organize those apps and add new apps as you find something else you like, and understand how to better navigate and interact with the little wonder on your wrist.

Update the Operating System

As of this writing, the most current version of the Apple Watch operating system is watchOS 8. Each new iteration of watchOS comes with a bevy of features that Apple is introducing, significantly upgrading, or updating.

I make sure that whatever device I'm using is running the latest version of the operating system and I think you should too. Doing so not only keeps you and your software hip and with it (as the kids say) but also helps you stay on top of problematic issues — such as bad folks who are out to harm you or your devices with malicious software or internet attacks — and prevents trouble from raising its ugly head when you're not looking.

You can update watchOS in two ways: using your iPhone or directly through your Apple Watch if it's already running at least watchOS 6.

REMEMBER

Regardless of which method you decide to use, your Apple Watch must be

>> **At least 50 percent charged.** As a matter of fact, just leave it on its charger during the entire update process; if the charge drops below 50 percent the update will stop.

>> **Connected to Wi-Fi.** If you're updating through your iPhone, it must be connected to Wi-Fi. If you're updating through your Apple Watch, it must be using Wi-Fi. Go to Settings⇨Wi-Fi on your watch to make sure it's connected.

Update with iPhone

One way to update watchOS is with an iPhone using the Watch app. Let's see how to get this party started.

A BIT MORE INFO ABOUT OPERATING SYSTEMS

An operating system (OS) is software that makes a computer or smart device (like your Apple Watch) work. The OS allows you to interact with your computer or smart device. You tell the OS that you want to perform a task by clicking, dragging, saying, or tapping something. Then the OS tells your device hardware what you want it to do, such as shut down, restart, play a music file, make a phone call, or what-have-you.

WARNING

Do not restart either your iPhone or Apple Watch during this process, and do not quit the Watch app. Also, do not remove your Apple Watch from its charger. Doing any of these things could cause problems with the update and bring on major headaches for you.

1. Open the Watch app on your iPhone and make sure you're looking at the My Watch tab. (If not, tap it at the bottom of the screen.)

   You must have the latest version of iOS installed on your iPhone to update watchOS with the Watch app.

TIP

   Keep your iPhone near your Apple Watch during the update process to make sure they maintain a good connection.

2. Go to General ⇨ Software Update.

3. If an update is available (as shown in **Figure 3-1**), tap to download and install it.

4. If prompted, enter your iPhone and Apple Watch passcodes.

   A progress wheel appears on the display of your Apple Watch — leave everything as is until the update completes! When the update is finished, your Apple Watch restarts.

Be patient: These updates can often take a while (even upwards to an hour, on occasion).

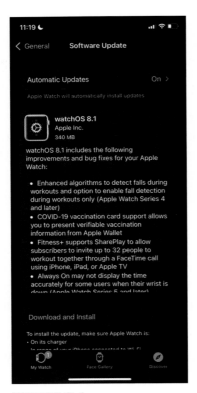

FIGURE 3-1

Update directly from your Apple Watch

Until watchOS 6, the only way to update watchOS was with an iPhone using the Watch app. Now, however, you can also update watchOS from your Apple Watch itself.

WARNING

Do not restart your Apple Watch or remove it from its charger during this process, and don't quit the Settings app. Doing so could cause problems with the update.

1. Open the Settings app on your Apple Watch.

REMEMBER

You must have at least watchOS 6 currently installed on your Apple Watch to update from the watch itself. Your Apple Watch must be connected to Wi-Fi to update directly with it.

2. Tap General⇨Software Update.

3. If an update is available (as shown in **Figure 3-2**), tap to download and install it.

FIGURE 3-2

4. If prompted, enter your Apple Watch passcode.

A progress wheel appears on the display of your Apple Watch — leave everything as is until the update completes! When the update is finished, your Apple Watch restarts.

Be patient: Updates sometimes go fairly quickly, but they usually take longer (even upwards of an hour or so).

Check Out the Apps

Every Apple Watch is endowed with a wealth of helpful (and often fun) apps that help you work and play. This section introduces you to the apps built into watchOS so that you'll know all the things you can do right out of the box. I also refer you to other chapters that focus on using particular apps.

Activity

The Activity app keeps track of daily movements. You've probably seen this app in action on some of Apple's commercials, or you've heard the phrase "Close your rings." That phrase refers to completing some or all of three daily tasks: move, exercise, and stand. The rings are multicolored and remind you throughout your day to get up and get around so that you stay active. Learn more in Chapter 11.

TIP

Check out Apple's website dedicated to closing your rings at www.apple.com/watch/close-your-rings/.

Alarms, Stopwatch, Timer, and World Clock

The names of the Alarms, Stopwatch, Timer, and World Clock apps speak for themselves. Check them out in Chapter 5.

Audiobooks

Chapter 13 tells you all about using the Audiobooks app to add audiobooks to your Apple Watch and listening to them on the go.

Blood Oxygen

The Blood Oxygen app is one of Apple Watch's crown jewels. The app uses sensors built into your Apple Watch Series 6 or later to measure the oxygen saturation of your blood. Chapter 10 shows you how to use this great new feature.

Calculator

I think the Calculator app speaks for itself, too.

Calendar

You'll be surprised at just how much you can do with the Calendar app on your watch. Chapter 7 provides the details.

Camera Remote

Did you know you can use your Apple Watch as a remote control for your iPhone's cameras? Very cool. See Chapter 12 for more info.

Compass

Lost? Well, with the Compass app, now you're found! Chapter 7 will guide you through this app.

Cycle Tracking

See Chapter 10 to find out how the Cycle Tracking app can help you keep track of your menstrual cycle.

ECG

No, ECG doesn't stand for *Egg Carton Garnishing* or *Eccentric Cat Groupies,* as disappointing as that may be. Instead, it stands for *electrocardiogram,* which you can now perform directly from your wrist. Now that I think about it, the ECG app is way cooler than the other two I mentioned. Chapter 10 has much more about this feature.

Find People

The Find People app is a great way to help you find Apple-using family and friends who you may be meeting or have simply been separated from during an outing. See Chapter 7 for more.

Heart Rate

Check your heart rate any time using the Heart Rate app and other tech built into your Apple Watch. Chapter 10 is where you find out how.

Home

If you're into home automation, the Home app will be right up your alley. You can control lights, thermostats, and more directly from your Apple Watch. See Chapter 15 for details.

Mail

Are you out on a run or walk and your Apple Watch alerts you to an important email you've been waiting for? Use the Mail app to read and reply to that email. Chapter 6 discusses this topic and more.

Maps

Chapter 7 covers how to use the Maps app on your Apple Watch to get step-by-step directions to your destinations, search for specific types of locations, and more.

Memoji

Memojis are emojis that you can personalize to look like you or someone (or something) else. The Memoji app, discussed in Chapter 6, helps you customize to your heart's content.

Messages

Send and receive text messages, as well as Digital Touch and voice messages, with the convenience of the Messages app on your Apple Watch. Chapter 6 covers this topic in detail.

Mindfulness

The Mindfulness app uses two functions, reflect and breathe, to help you stay centered and to remind you to take time to focus on your breathing. You'll find more in Chapter 10.

Music

With the Music app, you can add music to your Apple Watch and listen even if you're without your iPhone. Chapter 14 covers all the basics of this topic, as well as how to play your tunes from your wrist and add music from your iPhone's music library or Apple Music (if you're a subscriber).

News

Chapter 7 tells you how to stay up-to-date with the latest headlines using the News app on your Apple Watch.

Noise

The Noise app, which is discussed in Chapter 10, monitors the noise level in your surroundings and alerts you when things get noisy enough that your hearing could be detrimentally affected. We have four kids; you can imagine how much of a workout this app gets on a typical day in our home.

Now Playing

When you play audio by using the Music, Audiobooks, Books, Podcasts, and some third-party apps, you can control playback on your Apple Watch's display by using the Now Playing app. This being the case, this app is discussed in several spots throughout this book.

Phone

Yes, Virginia, you really can make and receive phone calls directly from your watch by using the Phone app. Chapter 6 handles this topic like a champ. ("Kirk to Spock. . . come in, Spock. . .")

Photos

When you're out for a stroll and you just can't stand the thought of passing by that stranger and not sharing pictures of your favorite people, Apple Watch has your back, even if you don't have your iPhone. The Photos app is discussed in Chapter 12.

Podcasts

If you're like me, you love a good podcast. The Podcasts app helps you synchronize specific podcasts from your iPhone to your Apple Watch so you can listen anywhere, even without your iPhone. See Chapter 14 for more.

Radio

Chapter 14 also helps you learn how to tune into stations based on genre, broadcast radio stations, and special stations curated for Apple Music subscribers by using the Radio app.

Reminders

The Reminders app is my constant companion, helping to keep me on track with this task and the next one. Create, respond to, and delete reminders from your Apple Watch with ease — Chapter 7 shows you how.

Remote

The handy little Remote app lets you control playback for music playing on your computer, whether you use a Mac or a Windows-based PC, as well as content playing on your Apple TV. Chapter 15 fills you in on the details.

Sleep

By wearing your Apple Watch to bed and using the Sleep app, you can get a view of your sleep habits. You can also create sleep schedules and set alarms. Chapter 10 gives you the rundown.

Stocks

The Stocks app is Gordon Gekko's favorite app, bar none. Need to keep track of your favorite stocks on the go? This app's for you, too. See Chapter 7 for the skinny.

Voice Memos

Sometimes a pen and paper just aren't handy. Chapter 7 shows you how to use the Voice Memos app to create voice memos using your Apple Watch. Never lose another great idea again!

Walkie-Talkie

The Walkie-Talkie app does just what its title implies: It lets you use your Apple Watch as a walkie-talkie. Read Chapter 6 so you can ditch the tin cans and string.

Wallet and Apple Pay

Your Apple Watch uses the Wallet app to safely store cards and passes you can use to purchase items or gain entrance to events (movies, concerts, and the like). Instead of carrying an actual wallet, read Chapter 9 to find out how to use Wallet and Apple Pay from your wrist.

Weather

No need to step outside to find out what's going on; get the weather report by raising your arm and taking a gander at the Weather app on your Apple Watch. Chapter 7 helps you stay on top of the weather in all its craziness.

Workout

The Workout app is one of the best apps Apple Watch has to offer, in my opinion. Using your watch to track your workouts has always been a core function of Apple Watch, and this version of the app is on point. Chapter 11 shows you how to track workouts, set goals, and introduces you to Apple Fitness+.

Organize Apps and Install New Ones

You can organize and rearrange apps not only on your iPhone but also on your Apple Watch. Two methods are available to customize the Apple Watch Home screen in any way you please.

Another trick you learn in this section is how to install third-party apps. Many apps you download for you iPhone have Apple Watch counterparts, which you can easily sync with your watch by using the Watch app. You can also download apps from the Apple Watch App Store.

Organize apps on your Home screen

The Home screen of the Apple Watch is similar to Home screens on your iPhone; it's the place you can go to see all installed apps. You can organize these apps directly on the watch or by using the Watch app.

To organize your Home screen via your Apple Watch:

1. Press the digital crown (the dial above the side button) to view your Apple Watch's Home screen.

2. Hold down on the app you want to move until all the apps start jiggling (yes, that's the term Apple uses to describe their motion), like mine are doing in **Figure 3-3**. Drag the app where you want it to reside and then let go to drop it in place.

3. Press the digital crown to make your apps stop the happy dance.

Your Apple Watch Home screen must be set to grid view to arrange apps in this way. If it's not in grid view, go to Settings⇨App View and tap Grid View. Apps are listed in alphabetical order if you use list view.

FIGURE 3-3

To organize your Home screen via your iPhone's Watch app:

1. Open the Watch app and navigate to My Watch ⇨ App View ⇨ Arrangement to see the Arrangement screen, shown in **Figure 3-4**.

FIGURE 3-4

2. Hold down on the app you want to move, drag it to its new location, and then let go to drop it in.

The Watch app on your iPhone automatically syncs with your Apple Watch to make the change.

Install new apps

The App Store on your Apple Watch is a great place to start discovering and installing new apps.

1. Press the digital crown to view your Apple Watch's Home screen, and then tap the App Store icon (refer to **Figure 3-5**).

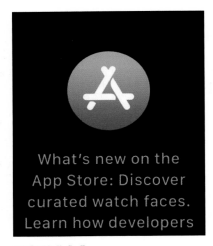

What's new on the App Store: Discover curated watch faces. Learn how developers

FIGURE 3-5

2. Swipe up and down on your display or rotate the digital crown to scroll through a list of suggestions. Or tap the Search field and enter the name of a particular app.

3. Tap an app to get more information.

4. When you find an app that you want to install, tap Get (if it's a free app) or the price (if the app isn't free), as shown in **Figure 3-6**.

5. Follow the prompts on the Apple Watch display to begin installation.

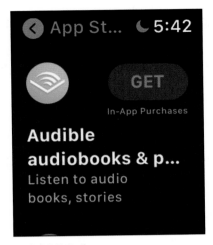

FIGURE 3-6

TIP

Apps you've previously downloaded to your Apple Watch but have since deleted display a download button, which looks like a cloud containing a downward-pointing arrow. If you currently have an app installed, you'll see an Open button.

When you download an app to your iPhone from its App Store, you can determine whether it also has a version for Apple Watch by scrolling to the Preview section and checking the small text under the pictures. After you install the app on your iPhone, by default it automatically installs its companion app on your Apple Watch. If you're like me and don't want apps to automatically be installed, you can disable this feature:

1. Open the Watch app on your iPhone.

2. If the My Watch tab isn't displayed, tap My Watch at the bottom of the screen, and then tap General.

3. Tap the Automatic App Install switch (refer to **Figure 3-7**) to toggle it off (gray).

FIGURE 3-7

Then, to manually install the Apple Watch version of an iPhone app by using the Watch app:

1. Open the Watch app and tap the My Watch tab at the bottom of the screen (if you're not already there).

2. Scroll down the page until you see the Available Apps section.

   You'll find a list of apps installed on your iPhone that also have companion Apple Watch apps (like the one in **Figure 3-8**).

3. Tap the Install button to the right of an app to install it on your Apple Watch.

   When the app has finished installing, it appears in the Installed on Apple Watch section of the My Watch tab.

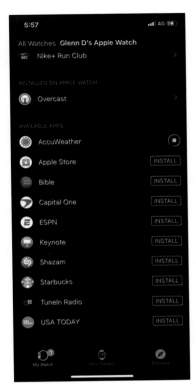

FIGURE 3-8

Know Your Status

Your Apple Watch displays tiny status icons and symbols designed to alert you of some happenings (such as notifications) or indicate that your Apple Watch is in a particular state (Do Not Disturb may be enabled, for instance). These icons and symbols typically appear at or near the top of the display.

Table 3-1 shows you some of the most common icons and symbols. Please refer to Apple's Support site at `https://support.apple.com/en-us/HT205550` for a comprehensive list.

TABLE 3-1 **Common Status Icons and Symbols**

Icon	What It Indicates
⚡	Apple Watch is in charging mode.
⚡	Low battery.
✈	Apple Watch is in airplane mode, which means all wireless connections (Bluetooth and Wi-Fi, for examples) are disabled.
🌙	Do not disturb mode is enabled, so you won't be alerted to notifications, calls, and the like.
🎭	Theater mode is enabled. This feature keeps your screen dark until you tap it or press the side button or digital crown.
📶	Wi-Fi is connected. Appears only when in Control Center (more on that later in the chapter).
📵	Your Apple Watch is not connected to your iPhone.
⬤	You've received a notification.
🔒	Apple Watch is locked; unlock it by using your passcode.
🏃	A workout is in progress. Tap to open the Workout app and see your workout stats.
ᵢₗₗ	Audio is playing on your Apple Watch. Tap the Now Playing symbol to view playback controls for the audio file you're playing.

Images in Table 3-1 courtesy of Apple, Inc.

Discover Control Center

Control Center is an easy-to-access one-stop-shop for making quick settings changes to your Apple Watch.

To open Control Center (shown in **Figure 3-9**), simply swipe up from the bottom edge of your Apple Watch's display. You won't see all the options on the screen at once, so swipe or rotate the digital crown to scroll down and see the rest. To close Control Center, just swipe down from the display's top edge.

FIGURE 3-9

TIP

You can also customize the items that appear in Control Center by clicking the Edit button (found at the bottom of Control Center). When the icons start jiggling, hold down on an icon and drag it to a new location, tap the red circle with a white – to delete an app, or tap the green circle with the white + to add an app. Tap Done when you're finished.

Table 3-2 shows you some of the most common of these icons and symbols. Please refer to Apple's Support site at `https://support.apple.com/en-us/HT205550` for a comprehensive list.

REMEMBER

Your Apple Watch is water resistant, not waterproof. In case you missed it in Table 3-2, please visit Apple's Support site at `https://support.apple.com/en-us/HT205000` to learn more about your Apple Watch's water resistance and exactly what you can and can't do with it in the water.

TABLE 3-2   **Common Status Icons and Symbols**

Icon	What It Does
100%	Displays remaining battery power.
((ᵗ))	Enables or disables cellular data (GPS + cellular models only).
✈	Enables or disables airplane mode.
🔕	Enables or disables silent mode, which prevents your watch from making any noises other than alarms.
🌙	Enables or disables do not disturb mode.
🔒	Indicates your Apple Watch is locked; tap the display to enter your passcode when you're ready to unlock it.
💧	Enables or disables the water lock feature. Please see `https://support.apple.com/en-us/HT205000` for more information on this feature and the water resistance of your Apple Watch.
((📱))	Pings your iPhone. Or hold down on the icon to ping and make your iPhone's light flash (which is helpful if it's dark where you're looking).
🔦	Turns on your watch's flashlight. Your display turns a bright white, providing some light. Turn off the flashlight by swiping down on your display.
🎭	Enables or disables theater mode.
🛜	Enables or disables Wi-Fi.
🔘	Connects or disconnects an audio device, such as an external Bluetooth speaker.
((📷))	Enables or disables Walkie-Talkie availability.
🛏	Enables or disables sleep tracking.

Images in Table 3-2 courtesy of Apple, Inc.

Learn All about the Dock

Open the dock, which is a navigational feature in watchOS, by pressing the side button. Then scroll through the list of apps (illustrated in **Figure 3-10**) by swiping up or down or rotating the digital crown. Tap an app to open it. Press the side button again to close the dock.

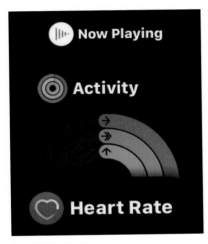

FIGURE 3-10

TIP

You can remove an app from the dock by swiping it to the left and tapping the remove button (a white X in a red square).

The dock grants you quick and easy access to either the most recent apps you've used or up to ten of your favorite apps — your call. You can determine which you see when you open the dock by using your iPhone's Watch app:

1. Open the Watch app.

2. If the My Watch tab isn't displayed, tap it at the bottom of the screen. Then tap Dock.

3. Tap either Recents or Favorites.

TIP

If you selected Recents, apps will be listed in the order they were opened.

You can customize your list of favorites by using your iPhone's Watch app:

1. Open the Watch app, go to My Watch ⇨ Dock, and tap Favorites to display a list of your installed apps.

   A default list is already populated under the Favorites section; others are found in the Do Not Include section.

2. Tap the Edit button in the upper-right corner.

3. To remove an app from the list, tap the red circle with the white – to the left of the app in the Favorites section. To add an app to the list, tap the green circle with the white + to the left of the app in the Do Not Include section (refer to **Figure 3-11**).

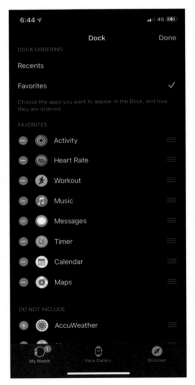

FIGURE 3-11

4. Reorder an app in the Favorites list by holding down on the three stacked white lines to the right of the app's name and dragging up or down.

5. When you're finished, tap Done in the upper-right corner.

Get In Touch with Haptics

Your Apple Watch can tap you on the wrist to notify or alert you by using a little tech called haptics, which combine software and hardware features, in this case vibration. Haptics are on by default for your Apple Watch, and can really come in handy, especially when you have your watch in silent mode. One of my favorite experiences with haptics is when my Apple Watch taps me on the wrist while I'm using turn-by-turn directions with the Maps app and am rapidly approaching a turn.

You can adjust the strength of these taps:

1. Open Settings on your Apple Watch.

2. Swipe or rotate the digital crown, and then tap Sounds & Haptics when you find it.

3. Swipe or rotate the digital crown until you see the Haptics section (shown in **Figure 3-12**), then tap either Default or Prominent to adjust the strength of haptic touches.

TIP You can also have your Apple Watch tap the time for you using the Taptic Time feature. To set it up, go to Settings⇨Clock⇨Taptic Time, toggle the switch on (green), and then tap Digits, Terse, or Morse Code to select how Taptic Time provides haptic feedback. Check out this Apple Support article for more details: `https://support.apple.com/guide/watch/adjust-brightness-text-size-sounds-haptics-apd62807a9f3/8.0/watchos/8.0`.

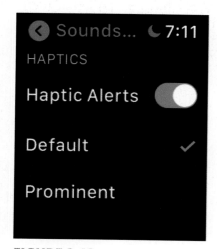

FIGURE 3-12

Chapter **4**

Making Your Apple Watch Accessible

'm always been impressed with Apple's commitment to accessibility. The idea that tools should work for anyone who wants to use them, no matter the physical obstacles they may face, is one that Apple has acknowledged since my earliest days as a fan. Accessibility is a fundamental right, and it's great to see company's like Apple (and others) prioritize that right so that almost anyone who wants to can use their devices.

This chapter explores the accessibility options watchOS affords users. It's my hope that you'll find something in the following pages to greatly enhance your Apple Watch experience.

TIP

You can set most of the upcoming accessibility options on the Apple Watch itself or in the Watch app on your iPhone. You may find that it's easier to use the Watch app and then tweak the options on the Apple Watch.

Brighter and Bigger

Sometimes, simply making things a bit brighter and a bit larger makes them easier to see. That is true especially for text on a screen the size of the one on your Apple Watch. Aside from making the Apple Watch the size of your iPhone (that would be a bit clunky to wear on your wrist, for sure), let's see how to adjust text brightness and size:

1. Open the Settings app on your Apple Watch.

2. Tap Display & Brightness.

3. Increase or decrease the brightness of your Apple Watch screen using the brightness controls (see **Figure 4-1**). You can either tap the left or right side of the controls, or tap Appearance and then rotate the digital crown.

4. Scroll down, and then tap Text Size. To increase or decrease the text size, tap the left or right side of the controls or rotate the digital crown (see **Figure 4-2**).

5. Tap < in the upper-left corner to return to the main Display & Brightness options.

6. To make the text heavier against the background, toggle the Bold Text switch on (green).

   If this doesn't suit your tastes, simply toggle the switch off.

FIGURE 4-1

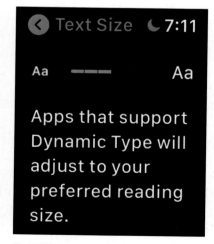

FIGURE 4-2

Let's Get Visual

watchOS includes many options to assist you with viewing items (text or otherwise) on your Apple Watch display. Sometimes you need more than just a brighter display, so let's see what else you can do on the visual front.

To explore additional visual settings:

1. Open the Settings app on your Apple Watch.

2. Tap Accessibility, and then toggle the switches shown in **Figure 4-3** on (green) or off for the following options:

   - *Bold Text* causes the text to be heavier. Sometimes the thin lines of certain characters may be difficult to see (especially against a dark background), so making the text bold increases the thickness of those lines.

   - *On/Off Labels,* on in **Figure 4-4**, displays a | or 0 next to a switch, in addition to other indicators, such as color.

   - *Grayscale* changes the color scheme of everything on your Apple Watch display to grayscale, removing all color.

   - *Reduce Transparency* helps make certain backgrounds on your Apple Watch more legible by increasing the contrast.

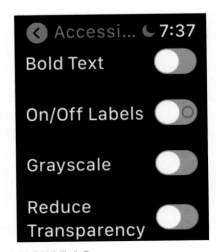

FIGURE 4-3

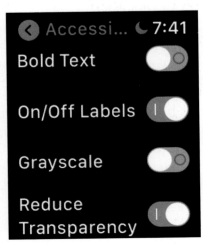

FIGURE 4-4

3. Tap Reduce Motion, which can cut down on some of the onscreen animation motions that may cause visibility issues. Then toggle the Reduce Motion switch on if this is a feature that could help you.

While in this section, decide whether to enable or disable the Auto-Play Message Effects option. When enabled, this option allows special effects (such as animations) in the Messages app to auto-play.

Another simple thing you can do to increase visibility is utilize the X-Large watch face, shown in **Figure 4-5**. For more on changing watch faces, please check out Chapter 5.

FIGURE 4-5

Zoom Zoom

No, you've not stumbled onto an old Mazda commercial in the middle of your book. Rather, in this section, I show you how to use the zoom feature to make viewing graphics and text on your Apple Watch display so much easier.

To begin, let's enable zoom:

1. Open the Settings app on your Apple Watch.

2. Tap Zoom.

3. Toggle the Zoom switch on (green).

Now that you have the feature enabled, it's time to use it! Limber up your index and middle fingers before proceeding; they'll get a good workout using zoom.

Here's what you can do with zoom:

» Using two fingers, double-tap the display to zoom in and again to zoom out.

» When zoomed in, use two fingers to pan to view other portions of the screen. You can also rotate the digital crown to pan. When you do, a little green zoom icon in the upper right shows you which portion of the screen you're viewing (seen in Figure **4-6**).

FIGURE 4-6

TIP

When panning, you can use the digital crown to scroll up and down the display, instead of using it to pan the display. Just tap once on the display — using two fingers — to toggle between the modes.

>> Using two fingers (told you you'd need to limber them up), double-tap the display, holding them down on the display on the second tap. Then, drag both fingers up or down the display to increase or decrease magnification, respectively.

TIP

Adjust the amount of magnification in the zoom function by going to Settings ⇨ Accessibility ⇨ Zoom, and then tapping – or + in the Maximum Zoom Level slider to decrease or increase the level, respectively.

Get Your Motor Runnin'

Let's be honest: The small display on the Apple Watch can be tough to navigate. Some folks have large fingers, while others may have other difficulties that prevent them from using the touchscreen and buttons effectively. watchOS includes some motor skill settings to help us out.

Touch Accommodations

First up is touch accommodations:

1. Open the Settings app on your Apple Watch.
2. Tap Accessibility, and then tap Touch Accommodations.
3. Toggle the Touch Accommodations switch on (green), and then tap OK in the confirmation message.

With touch accommodations enabled, you can utilize three options:

>> *Hold Duration* allows you to adjust the time it takes for Apple Watch to recognize a touch. Toggle this switch on (green), and then tap the white – or + button (shown in **Figure 4-7**) to decrease or increase the time, respectively.

>> *Ignore Repeat* allows you to adjust the time it takes for Apple Watch to recognize multiple touches as a single touch. Toggle

this switch on (green), and then tap the – or + button to decrease or increase the time, respectively.

» The options in the *Tap Assistance* section allow you to tell your Apple Watch whether to recognize the initial contact you make with the display or the final contact you make as the location of where the touch took place. Select between Off, Use Initial Touch Location, or Use Final Touch Location, illustrated in **Figure 4-8**.

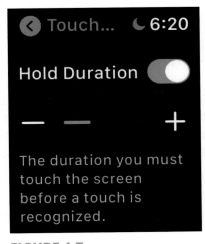

FIGURE 4-7

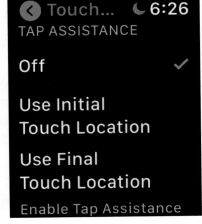

FIGURE 4-8

Adjust the click speed

Next, let's adjust the click speed for when you need to click the side button. The side button is small, so clicking it in a timely manner can be a bit frustrating when, for example, you're using it to activate Apple Pay. To adjust that speed:

1. Open the Settings app on your Apple Watch.

2. Tap Accessibility, and then tap Side Button Click Speed.

3. Select Default, Slow, or Slowest to make the adjustment, and then tap < in the upper-left corner to back out of the option.

Tap Out

Using haptic feedback, your Apple Watch can help you tap out if you're in a wrestling or UFC match and your opponent has you in quite the precarious and inescapable predicament.

Okay, not really. Sorry to disappoint all the wrestlers and UFC fighters in the audience. Actually, this section's title refers to tapping out the time on your wrist. Apple Watch uses its haptic technology to tap out various long and short taps, representing hours and even minutes. This option is called taptic time.

Taptic time uses one of three modes to provide those timely taps:

» **Digits:** The tap sequence begins with long taps that signify every 10 hours, and then short taps for every subsequent hour, and then long taps again to signify each span of 10 minutes, followed by more short taps signifying each additional minute.

» **Terse:** The tap sequence starts with long taps that signify 5 hours each, and then short taps for every subsequent hour, and then more long taps that each signify spans of 15 minutes.

» **Morse Code:** You'll want to know Morse Code for this one because your Apple Watch will tap out each digit of the time using it.

REMEMBER

If you have speak time enabled in Settings ➪ Clock, taptic time won't work unless you use the Control with Silent Mode option and have your Apple Watch in silent mode. You can check to see your Apple Watch is in silent mode by swiping up from the bottom of the screen while viewing the watch face to open Control Center.

To set up taptic time:

1. Open the Settings app on your Apple Watch.

2. Tap Clock, and then tap Taptic Time (boy, that's a bit of a tongue twister).

3. Toggle the switch on (green), and then tap Digits, Terse, or Morse Code, as shown in **Figure 4-9**.

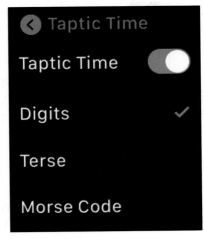

FIGURE 4-9

4. To feel the time, place and hold two fingers on the Apple Watch display (while viewing the watch face).

   You'll feel the taps corresponding to the mode you selected in Step 3.

TIP

You really — and I mean, *really* — have to pay attention to the sequence of taps to discern the correct time with haptics. Focus is the name of the game, along with remembering which mode you've chosen for the taps.

You Speak to Me, Apple Watch!

VoiceOver is an incredibly useful screen-reading tool that will allow your Apple Watch to tell you what's on its display screen. If you're having difficulty seeing the options on the screen, Apple Watch is happy to tell you what's there.

When you have VoiceOver enabled, every item on the display is selectable, and once it's selected, VoiceOver reads what it is to you. You can tell when an item is selected by the white line that encompasses the item, whether it be a button, a word, or a graphic on the screen (such as your watch face). You can use a series of gestures to select items on the screen, or use the digital crown to scroll through all the selectable items.

TIP

VoiceOver is a wonderful feature, but be warned that it takes lots of practice to learn how to successfully navigate your Apple Watch on a regular basis with it. Don't be discouraged if it takes a while to remember all the various gestures.

Set VoiceOver options

Instead of enabling VoiceOver and then worrying about its various options, I think it's a good idea to start with the options first. VoiceOver is such a different way to navigate Apple Watch that enabling it and then trying to use it to make setting adjustments makes the learning curve steeper.

Let's first get to VoiceOver:

1. Open the Settings app on your Apple Watch.

2. Tap Accessibility, and then tap VoiceOver.

Now let's explore the options:

» *Speaking Rate:* Tap the turtle or the hare button (shown in **Figure 4-10**) to decrease or increase VoiceOver's rate of speech, respectively.

» *VoiceOver Volume:* Tap the button (refer to Figure 4-10) on the left or right to decrease or increase VoiceOver's volume, respectively.

» *Haptics:* Toggle this switch on (green) to allow haptic feedback while using VoiceOver.

» *Siri Voice:* Toggle this switch on (green) to have VoiceOver use the same voice as Siri.

» *Braille:* Tap to wirelessly connect a braille display to your Apple Watch via Bluetooth, as well as make adjustments to how they interact.

USING BRAILLE DISPLAYS WITH VOICEOVER

Apple has a couple of great articles on their Support site to assist with braille displays. For a list of braille displays supported by Apple Watch, visit `https://support.apple.com/en-us/HT211966`. Also, check out `https://support.apple.com/en-us/HT211967` to learn braille commands to use with VoiceOver.

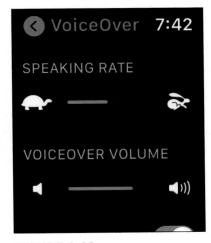

FIGURE 4-10

» *Speak Hints:* VoiceOver can give you hints about how to engage an item on the display. For example, depending on what you may have selected, it may tell you to double-tap or swipe the item. Toggle this switch on or off to engage or disengage the feature.

» *Digital Crown Navigation:* Instead of swiping around and tapping items to select them, enabling this option allows you to rotate the digital crown to make selections.

TIP

Digital crown navigation can be a bit flaky. For instance, sometimes I can get stuck on an item, and no matter how much I rotate the darned digital crown, I don't go anywhere. To get out of that rut, I ask Siri to turn off VoiceOver, and then turn it back on again.

» *Speak on Wrist Raise:* Enable this option to hear the time every time you raise your wrist.

» *Screen Curtain:* This feature turns off your Apple Watch display for privacy. The only way to interact with your watch is through VoiceOver.

⚠ WARNING

The only way to disable Screen Curtain is to open the Apple Watch app on your iPhone, go to the My Watch tab, and choose Accessibility. Then toggle the Screen Curtain switch to off (gray).

» *Speak Seconds:* When seconds are displayed on the screen, VoiceOver will read them aloud.

» *Rotor Languages:* The rotor in VoiceOver is a dial that gives you quick access to several VoiceOver options. To use the rotor, place two fingers on the display and then turn them clockwise or counterclockwise to view the options onscreen (as shown in **Figure 4-11**) and to hear them spoken by VoiceOver. To adjust one of the options, swipe up or down on the screen when you hear or see the one you want. The Rotor Languages option lets you add more than one language to the rotor.

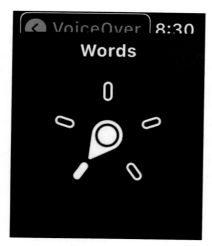

FIGURE 4-11

Enable VoiceOver

To enable VoiceOver:

1. Open the Settings app on your Apple Watch.

2. Tap Accessibility, and then tap VoiceOver.

3. Toggle the VoiceOver switch on (green).

 WARNING All apps that are native to or that come with watchOS are designed to work with VoiceOver. However, note that not all third-party apps work with VoiceOver.

Navigate with VoiceOver

Getting around your Apple Watch with VoiceOver will take some getting used to, but a little practice will go a long way.

Here's how to navigate with VoiceOver:

» Move around your Apple Watch display by moving your finger around it or by tapping an item. As you touch an item, its name (and sometimes its description) are spoken to you.

» You can swipe left or right and up or down to move to adjacent items and to move to new pages.

» Normally, a single tap will open an item, toggle a switch, and so on. With VoiceOver on, anything you'd normally do with a single tap now requires a double-tap. For example, to toggle a switch on or off, select it and then double-tap.

» You'll quickly notice that VoiceOver can be quite chatty. If VoiceOver is in the middle of reading information from the screen and you'd like it to take a break, simply tap once on the screen with two fingers. When you're ready for VoiceOver to start reading again, just tap the screen again with two fingers.

» If you need VoiceOver to speak a bit louder or quieter, adjust the volume by double-tapping the display with two fingers, being sure to hold down both fingers on the display as part of the second tap. Then slide your two fingers up or down the display to increase or decrease volume, respectively.

If this process turns out to be challenging, you can also open the Apple Watch app on your iPhone, go to My Watch ⇨ Accessibility ⇨

VoiceOver, and then drag the VoiceOver Volume slider, as shown in **Figure 4-12**.

FIGURE 4-12

» Sometimes an item may allow different actions to be taken with it. When this happens, VoiceOver will say, "Actions available." Swipe up or down on the display to select an action, and then double-tap the display with one finger to perform that action.

» If you've gone into menus or options that you didn't intend to, you can use a technique Apple calls a two-finger scrub to back up one level at a time in the menu structure. To perform the scrub, use two fingers to swipe left to right, right to left, and then left to right again, while keeping your fingers on the display the entire time. This is essentially like tracing the letter Z with two fingers.

» It may be easier on some screens to rotate the digital crown to select items while using VoiceOver. To enable this option from any screen, triple-tap the display using two fingers. You should hear "Digital crown navigation on" if you successfully enabled it. Now simply rotate the digital crown, and it will move from item to item on the screen. For example, in **Figure 4-13**, the workout complication is selected, and double-tapping anywhere on the display will launch the Workout app. To disable digital crown navigation, just triple-tap the display with two fingers again.

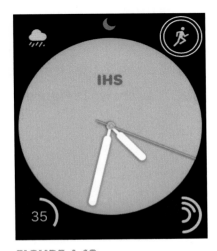

FIGURE 4-13

TIP

If you get stuck somewhere and need to turn off VoiceOver, get Siri to do it for you. Simply say, "Hey, Siri, turn off VoiceOver" and it will be done.

Hear Ye, Hear Ye!

Apple is also keen to assist Apple Watch fans who may have difficulties with audio. Several watchOS features are at your disposal.

Mono audio

When using headphones, stereo audio can play havoc with someone who may be hard of hearing or deaf in one ear. Since stereo audio consists of two separate audio tracks, one right and one left, some or all of the audio in one of the channels may be missed. Imagine listening to a conversation and hearing only some of the words.

Apple Watch's Mono Audio setting forces both audio channels to play in both the right and left speakers of the headphones.

To enable and adjust the Mono Audio setting:

1. Open the Settings app on your Apple Watch.

2. Tap Accessibility, and then swipe or scroll down to the Hearing section.

3. Toggle the Mono Audio switch on (green), as shown in **Figure 4-14**.

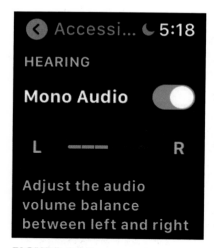

FIGURE 4-14

4. If you want to adjust the audio balance, increasing the volume in either the left or right speaker, tap L or R in the area under the Mono Audio switch (refer to Figure 4-14).

   You can adjust the audio balance even when Mono Audio is off.

Adjust AirPods options

If you're the lucky owner of a pair of AirPods or AirPods Pro, you can also customize accessibility options for them through your Apple Watch.

To adjust these options:

1. Open the Settings app on your Apple Watch.

2. Tap Accessibility, and then tap AirPods.

3. From here, you can customize or enable three options:

   - *Press Speed:* Tap Default, Slower, or Slowest (see **Figure 4-15**) to adjust the rate of speed needed to press the button on your AirPods two or three times.

   - *Press and Hold Duration:* Tap Default, Shorter, or Shortest to adjust the amount of time needed to press and hold down the button on your AirPods.

   - *Noise Control:* Toggle the Noise Cancellation with One AirPod switch on (green), as shown in **Figure 4-16**, to enable noise cancellation (if your AirPods support the feature) when only one AirPod is being used.

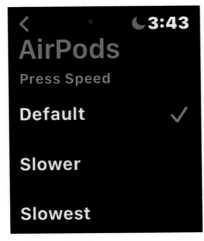

FIGURE 4-15

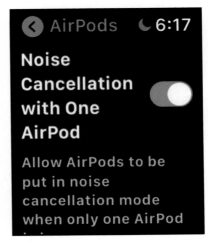

FIGURE 4-16

Finally, the Accessibility Shortcut

watchOS includes a nifty accessibility shortcut feature that allows you to enable one of three possible accessibility options by simply triple-clicking the digital crown.

To enable and use the accessibility shortcut:

1. Open the Settings app on your Apple Watch.

2. Tap Accessibility, and then tap Accessibility Shortcut (you'll need to swipe or scroll to the bottom of the page to find it).

3. Tap to choose which feature to enable with a triple-click of the digital crown: VoiceOver, Zoom, or Touch Accommodations (shown in **Figure 4-17**).

4. Triple-click the digital crown to enable the selected feature, and triple-click it again to disable the feature.

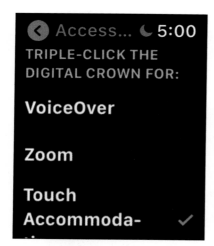

FIGURE 4-17

2
Beginning to Use Your Apple Watch

IN THIS PART . . .

Customizing watch faces

Communicating using Apple Watch

Keeping organized with Apple Watch

Getting to know Siri

Using Apple Pay and Wallet

IN THIS CHAPTER

» Discover watch faces and complications

» Set clock options

» Keep pace with Stopwatch and Timer

» Set alarms

» Check the time around the globe

Chapter **5**

What Time Is It?

An Apple Watch would be still be pretty great even if it didn't tell the time, but then it'd have to be called something else. Perhaps the Apple Wrist Wearable Thingy, or something cool and catchy like that. (It's no wonder I don't work for Apple's marketing group, right?)

Well, thankfully Tim Cook and the gang made the wise decision in 2015 to include a clock feature in their fancy new wearable device, so they were able to give it its now well-known moniker without looking really silly. But not only did they include a clock app, they made it exceedingly customizable so that everyone who wore one could make it suit their tastes and lifestyle. The first Apple Watch offered only 10 watch faces, but that number has bloomed to more than 40! Trust me when I say that there's a watch face for just about anyone.

In addition to telling the time, Apple Watch includes other time-centric apps such as Stopwatch, Timer, Alarms, and World Clock. You explore them all in this chapter, so get going!

Face-to-Face

Your watch face is what you see by default every time you look at your Apple Watch display. Thankfully, you can select from a multitude of watch faces so that your Apple Watch can best reflect your tastes and needs. Some faces are bare bones while others resemble an airplane's instrument panel, and most are customizable (in some cases, to a great extent).

As mentioned, watchOS now sports more than 40 watch faces for you to choose from. Examples include the following:

» Artist

» Chronograph Pro

» Explorer

» Fire and Water

» Memoji, shown in **Figure 5-1**

» Modular Compact

» Numerals Duo

» Photos

» Solar Dial

» Toy Story — that's right, Buzz, Woody, and the gang on your Apple Watch (see **Figure 5-2**)!

And many more!

To see a full list of the latest watchOS watch faces, including detailed descriptions, check out the article entitled *Apple Watch Faces and Their Features* at Apple's Support site: https://support.apple.com/guide/watch/faces-and-features-apde9218b440/watchos.

FIGURE 5-1

FIGURE 5-2

COMPLICATING MATTERS

We've all seen the tiny dials, hands, faces, and other features on typical watches; these little extras (or *special features*, as Apple refers to them) are called *complications,* and they're basically any function on the watch face that isn't related to telling time. An example is the date dial or the subdials on a chronograph. The faces on an Apple Watch are no different. Some watch faces allow you to choose from more than 50 complications! Mind you, you can't fit them all on the display at once, but you do have a hefty selection to choose from for the space available.

Examples of complications include the following:

- Activity
- Battery
- Calendar
- Date
- Heart rate
- Weather

Choose and Customize Faces

Now that you're aware of the cornucopia of watch faces at your disposal, allow me to show you how to choose the one that suits your style. Then you take a look at how to further customize things.

Change your watch face

To quickly select a watch face:

1. Raise your wrist to view the current watch face.

2. Scroll through the default selection list by swiping a finger from one edge of the screen to the other, either left to right or right to left.

3. Stop on the face you want to use. Done!

Add a face to the selection list

If you don't find what you're looking for on the default selection list, you can add other watch faces to the list:

1. While viewing the current watch face, touch and hold down on the display until the Edit screen appears.

2. Swipe from right to left on the display until you see the New (or add) button, as shown in **Figure 5-3**, and tap it.

3. Rotate the digital crown until you see a watch face you'd like to use, and then tap to select it.

Remove a face from the selection list

You can remove a face in your selection list as easily as you can add one. This feature comes in handy when your list is a mile long and it takes 10 minutes to scroll through.

1. While viewing the current watch face, touch and hold down on the display until the Edit screen appears.

FIGURE 5-3

2. Swipe up from the bottom of the screen to see the Remove button, as shown in **Figure 5-4**.

3. Tap the Remove button, and then tap a new face from the list to use.

FIGURE 5-4

REMEMBER

No worries — removing a face from the list doesn't delete it from your Apple Watch; it's just removed from the selection list. You can always add it back later.

Customize your watch face

In this section, you customize your watch face to meet your likes and needs. To customize any watch face:

1. While viewing the watch face, touch and hold down on the display until the Edit screen appears.

   In this example, I'm using the California watch face, but feel free to follow along with the watch face of your choice.

2. Tap the Edit button (see **Figure 5-5**).

   Depending on the watch face, you'll have several groups of settings you can customize. The name of the settings group you're currently viewing is shown at the top of the display, as shown in **Figure 5-6**.

FIGURE 5-5

FIGURE 5-6

3. To change the settings group, swipe left or right on the display.

   A thin green line outlines the setting or complication you're editing.

4. To customize complications:

   a. Swipe to the Complications group, as shown in **Figure 5-7**.

   b. Tap the complication you want to edit.

c. Rotate the digital crown to select a different complication from the list, as shown in **Figure 5-8**.

d. Tap the new complication to add it to your watch face.

FIGURE 5-7

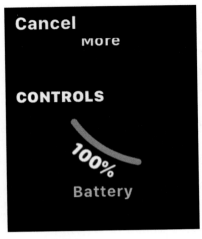

FIGURE 5-8

5. When you've finished editing your watch face, press the digital crown to save the settings and view the new look.

   Figure 5-9 shows the customizations I've made to the original California watch face (refer to Figure 5-5 for comparison).

Meet the Face Gallery

The Face Gallery displays all available watch faces — and several ready-made customizations — in one convenient and easy-to-view location. The Face Gallery is in the Apple Watch app on your iPhone.

To check out and use the Face Gallery:

1. Open the Apple Watch app on your iPhone and tap the Face Gallery tab at the bottom of the screen.

2. To view the various categories of watch faces, swipe up and down.

   The Infograph, Infograph Modular, and Kaleidoscope faces are shown in **Figure 5-10**.

FIGURE 5-9

3. When you get to a category you'd like to explore, swipe it left or right to view various iterations of it.

4. Tap a watch face to view its configuration settings, shown in **Figure 5-11**.

5. Peruse the various settings you can customize, and tap settings groups such as Color or Complications to view the available options and make any desired changes.

6. When you're finished editing, tap the Add button under the watch face's name (refer to Figure 5-11) to make it your new face and add it to your Apple Watch's selection list.

Share a watch face

Do you have a customized watch face you're particularly pleased with and want to share with friends? Easily done!

1. View the watch face you'd like to share on your Apple Watch display.

2. Tap and hold down on the display to view the Edit screen.

FIGURE 5-10

FIGURE 5-11

3. Tap the share icon to the left of the Edit button (see **Figure 5-12**).

   The share icon looks like a box with an arrow pointing up.

4. Tap either the Messages or Mail button. Then on the New Message screen (see **Figure 5-13**):

   a. Tap the Add Contact button.

   b. Select the contact(s) you want to share with.

   c. Add a message, if you like.

   d. Tap the Send button (scroll down to see it).

FIGURE 5-12

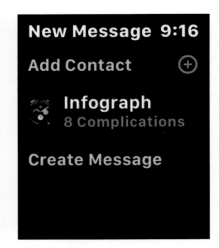

FIGURE 5-13

Discover Clock Options and Settings

You can customize how the Clock app works by using the Settings app. Several options allow you to bend the app to your will.

To access the Clock app settings:

1. To access the Home screen, press the digital crown once (maybe more, depending on where you are or which app you're in).

2. Tap to open the Settings app on your Apple Watch.

3. Tap Clock to open its settings.

Now, let's explore what these settings are and what they can do for you:

>> **Set Watch Display Time Ahead:** Some folks like to set their clocks a few minutes ahead of the actual time to help them stay on time for appointments and other events. This settings allows you to select how far in advance of the actual time you want to set the displayed time to on your Apple Watch.

1. Tap the +0 min button to adjust the time ahead in minutes.

2. Rotate the digital crown until you reach the desired amount of time ahead minutes, as shown in **Figure 5-14**.

3. Tap the Set button.

4. When you return to the Clock settings, note the actual time at the top of the screen compared to the time ahead that you've now set for your Apple Watch display (see **Figure 5-15**).

FIGURE 5-14

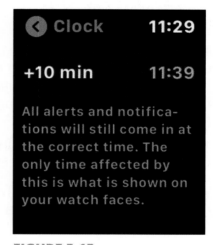

FIGURE 5-15

REMEMBER

Alerts and notifications still occur at their correct times; only the time on your Apple Watch display is changed by this setting. For example, suppose you set this setting to +10 minutes; if the actual time is 9:30, your Apple Watch will display 9:40. If you have an alarm set for 10:00, the alarm will indeed occur in real time at 10:00, but the display on your Apple Watch will show 10:10.

» **Chime:** Toggle this switch on (green) to have your Apple Watch play chimes at the top of every hour.

» **Sounds:** Tap to select whether Bells or Birds play when Chimes is enabled.

» **Speak Time:** Toggle this switch on (green) to have Apple Watch speak the time to you when you hold two fingers on the watch face.

» **Control with Silent Mode:** Tap to enable this option (see **Figure 5-16**), which prevents the time from being spoken when silent mode is turned on in Control Center. This option appears only when Speak Time is enabled.

» **Always Speak:** Tap to enable this option (refer to Figure 5-16), which allows the time to be spoken even when silent mode in turned on in Control Center. This option appears only when Speak Time is enabled.

» **Taptic Time:** Tap this option to access the toggle switch to enable it, and to select which method of haptic feedback is used to deliver the time (Digits, Terse, or Morse Code; refer to the section on tapping in Chapter 4 for more info), as shown in **Figure 5-17**.

This option is not available when Always Speak is enabled for the Speak Time option.

» **Watch Face Notifications:** Tap this option to be notified when a new watch face becomes available for your Apple Watch.

TIP

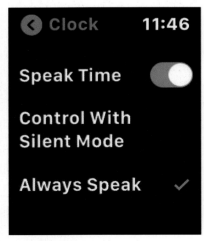

FIGURE 5-16

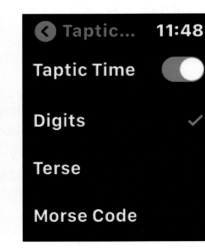

FIGURE 5-17

>> **Monogram:** Tap this option to select up to five characters to display in the monogram complication.

The monogram complication is available only for the California, Color, Infograph, Meridian, and Typograph watch faces.

>> **Siri Face Data Sources:** Tap this option to allow apps on your Apple Watch to provide information to the Siri watch face. Simply toggle the switch of an app to on (green) to utilize it as a resource.

REMEMBER

Use Stopwatch and Timer

The Stopwatch and Timer apps on your Apple Watch are extremely useful when their services are needed, and they're just as extremely simple to use.

Stopwatch

The Stopwatch app keeps track of the elapsed time in hours, minutes, and seconds, just like its analog counterpart. If you need to time how long it takes to run a mile or hold your breath, Stopwatch is there to do your bidding.

1. Access the Home screen by pressing the digital crown at least once (depending on where you are or what app you're already using on your Apple Watch).

2. Tap the Stopwatch app icon to open it.

3. To begin timing an event or action, tap the green Start button.

4. To record the time of specific portions of an event or action, while keeping the original time for the total event or action, tap the Lap button.

   To record multiple laps, as shown in **Figure 5-18**, press the Lap button multiple times.

5. To stop keeping the time of an event or action, tap the red Stop button (refer to Figure 5-18).

6. To begin keeping the time from where you last left off, tap the green Start button again. Or to start over, press the Reset button.

You can change the format of the Stopwatch app between Digital (default), Analog, Graph, or Hybrid (see **Figure 5-19**) by tapping the watch display when the Stopwatch app is open.

FIGURE 5-18

FIGURE 5-19

Want a record of an event's or action's times? Take a screenshot. Simply press the side button and the digital crown at the same time. The screenshot image is saved in the Photos app on your iPhone. (You'll find it in Albums ➪ Screenshots.)

Timer

Use the Timer app to set a desired length of time to complete an event or action.

1. To access the Home screen, press the digital crown at least once (depending on where you are or what app you're already using on your Apple Watch).

2. Tap the Timer app icon to open it.

3. To quickly begin a timer for an event or action, tap a predetermined duration.

4. Scroll to the top of the app screen and tap the Custom button to set a custom timer duration:

   a. Tap the Hours, Minutes, or Seconds field, as shown in **Figure 5-20**.

   b. To set the desired amount of time for each field, swipe up or down, or rotate the digital crown.

   c. To begin the timer, tap the green Start button.

5. To pause the timer, tap the orange Pause button in the lower-right corner (see **Figure 5-21**). To resume, tap the orange Play button. To shoo the timer away, tap the Cancel button (white X) in the lower-left corner.

FIGURE 5-20

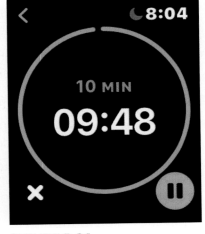

FIGURE 5-21

Sound the Alarm(s)!

Need help waking up? Want to be notified of a certain event at a specific time? Would you like to be reminded when it's time to quit playing video games? Set an alarm, and you're good to go.

1. Access the Home screen by pressing the digital crown at least once (depending on where you are or what app you're already using on your Apple Watch).

2. Tap the Alarms app icon to open it.

3. Tap the orange Add Alarm button to create a new alarm.

   If you've created several alarms, you may need to swipe down or rotate the digital crown to find the Add Alarm button.

4. Tap the hour or minute field, and then rotate the digital crown to make a selection (see **Figure 5-22**).

   An orange dot helps indicates the selected time on the clock.

5. Tap the AM or PM button.

WARNING

   Be sure to get the AM or PM setting correct! Goofing up this one can cause serious issues if the alarm is set for an important event.

6. To create and enable the alarm, tap the green Set button.

7. To view other options for an alarm (see **Figure 5-23**), tap it in the list:

   - Tap the Change Time button to adjust the time for the alarm.

   - Tap the Repeat button to set the alarm to repeat at certain intervals, such as every day, weekdays, weekends, or for specific days.

   - Tap the Label button to name the alarm.

   - Toggle the Snooze switch on (green) or off to allow you to enable or disable this option. The duration of the snooze is 9 minutes and can't be changed. When the alarm begins, you can tap the Snooze or Stop button on the display, or you can press the digital crown to snooze or the side button to stop the alarm.

   - Tap the Delete button to remove the alarm from your Apple Watch.

8. Toggle the switch of an alarm on (green) or off to enable or disable it.

TIP

You can also use your Apple Watch as a nightstand clock! To enable this option on your Apple Watch, go to Settings ⇨ General ⇨ nightstand mode, and then toggle the switch to on (green). When ready to use your Apple Watch in nightstand mode, connect it to its charger. The current time will appear when you touch the display, along with the time of any alarm you have set.

FIGURE 5-22

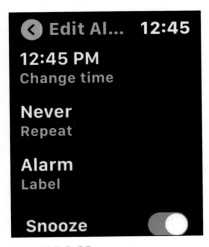

FIGURE 5-23

What Time Do You Have, Mate?
Or Amigo? Or Vinur? . . .

The World Clock app is a neat way to quickly see what time it is in other parts of the world. This approach certainly beats counting the hours in your head based on your current time zone, assuming, of course, that you know how many hours ahead or behind you the location you're curious about is.

It's easy to add cities, view more detailed information about them, and remove them from your World Clock list:

1. Access the Home screen by pressing the digital crown at least once (depending on where you are or what app you're already using on your Apple Watch).

2. Tap the World Clock app icon to open it.

3. Tap the orange Add City button to add a city to your list.

   If multiple cities are listed, you may need to swipe down or rotate the digital crown to find the Add City button.

4. Use scribble or dictation to enter the name of the city you're looking for.

Scribble appears by default, but you can easily switch to dictation by tapping the dictation icon (microphone) in the lower-right corner. Switch back to scribble if you like by tapping the scribble icon (a finger tracing a circle) in the lower right.

5. When presented with a lists of cities based on your search criteria, scroll through the list (by rotating the digital crown or swiping up or down on the screen) and tap the name of the city you want to add.

   When viewing your cities in World Clock (as I'm doing in **Figure 5-24**), you'll see their current time, as well as how many hours ahead or behind they are from your current location.

6. Tap a city name to see more detailed information for it (see **Figure 5-25**), including a map of its location and times for sunrise and sunset.

TIP

When viewing detailed information, use the digital crown to rotate through your list of cities.

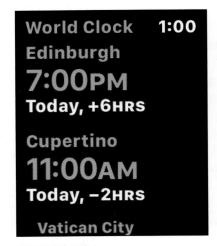

FIGURE 5-24

FIGURE 5-25

» **Send messages to friends and family**

» **Read and reply to email**

» **Chat with Walkie-Talkie**

Chapter **6**

Communicating with Apple Watch

I still vividly remember the first time I took a phone call on my Apple Watch. I was filling up the tank at a gasoline station when my wife called, and the little black device on my left wrist vibrated and sounded like a phone ringing. I was surprised because I wasn't expecting a call. But I wasn't as surprised as the person on the other side of the pump, who just stood there and stared, mouth wide open, as a voice emanated from my wrist and I spoke back. After a few seconds, the call ended, but the person continued to stand there and stare. I just smiled as I got into my car and drove away while — no kidding — the staring continued, following my car as I left the premises. To be fair, this was right before the Apple Watch went on sale; I'd received one early from Apple to write about the new toy, so they weren't ubiquitous like they are today. Good times, good times.

These days, whenever I hold up my wrist to take a phone call on my Apple Watch, I momentarily imagine I'm in a spaceport somewhere answering the call of my space squadron commander. Or that I'm Dick Tracy tracking down a sinister villain. Or maybe (and frankly, by far the coolest) I'm Bond — James Bond (and while I'm at it, let's go with the Daniel Craig incarnation) — answering an urgent call from M on one of Q's newest gadgets.

Seriously, answering calls on your wristwatch is something literally out of the comics or sci-fi. And yet, millions of people use their Watch to communicate by not only calling but also by text message and email. In this chapter, I cover those topics and more as you learn to communicate with the world using your Apple Watch.

Calling All Apple Watch Users!

Have I mentioned my fascination with making calls from my Apple Watch yet? The readership answers with an emphatic "Yes!" so I'll move on.

Time to start making calls with this little wonder on your wrist. No more dillying or dallying, just jump in.

REMEMBER

In most cases, if you don't have an Apple Watch GPS + cellular model (one that supports a cellular connection separate from your iPhone), you'll need to keep your iPhone close by to make and receive calls. The one exception is if you use Wi-Fi Calling, as discussed in the very next section. See Chapter 1 for more information on GPS + cellular models, or visit `www.apple.com/watch/cellular` to read Apple's whole take on these models and which carriers support them.

Call with Wi-Fi? Oh, my!

Some cellular carriers allow you to make phone calls from your iPhone or Apple Watch by using a Wi-Fi connection. You can even use your Apple Watch without its paired iPhone, as long as the Watch is connected to a Wi-Fi that your iPhone has been connected to at some point in the past. This feature comes in handy if your cellular connection isn't all that great (or even non-existent) in your current location.

TIP

Check Apple's Support site here for a list of carriers in your part of the world and to find out if they support Wi-Fi Calling: `https://support.apple.com/en-us/HT204039`.

If your carrier does support Wi-Fi Calling, you'll want to enable it to allow your iPhone and your Apple Watch to use the feature. To do so:

1. Open the Settings app on your iPhone.

2. Tap Cellular, and then tap Wi-Fi Calling in the section named after your carrier, as shown in **Figure 6-1**. In this example, my carrier is AT&T.

3. Toggle the Wi-Fi Calling on This iPhone switch on (green), as shown in **Figure 6-2**.

4. Toggle the Add Wi-Fi Calling for Other Devices switch on to allow your Apple Watch and other devices you use with your iCloud account (such as an iPad, perhaps) to utilize the feature.

FIGURE 6-1

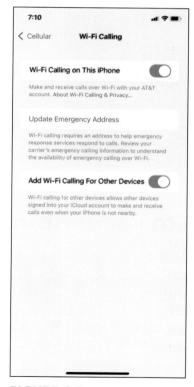

FIGURE 6-2

Good to go with Wi-Fi Calling on your Apple Watch! Now, if you try to make a call with Apple Watch but it is not near your iPhone but is connected to a familiar Wi-Fi, you're all set.

Answer a call

If you've not changed the default settings on your Apple Watch, you'll definitely know when a call is coming in. Your Apple Watch will use both a ringtone and a haptic (several taps on your wrist) to let you know someone wants to talk. You'll also see their name or phone number on your Apple Watch display, along with text saying *Incoming Call*, as shown in **Figure 6-3**.

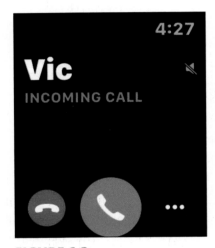

FIGURE 6-3

When you see, hear, or feel that call coming in, you have some options at your disposal:

» To answer, simply tap the green answer button (refer to Figure 6-3) and start talking.

» To hang up and send the call to voicemail, tap the red decline button (refer to Figure 6-3).

» To see additional options, tap the more button (three white dots). Then

- To place the call on hold until you can answer it with your iPhone, tap the Answer on iPhone button, shown in **Figure 6-4**. The person will hear a repeated tone while waiting.

- To send a message to the caller instead, swipe up (or rotate the digital crown) to scroll down the screen and tap one of the options in the Send a Message section (see **Figure 6-5**).

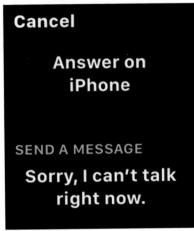

FIGURE 6-4

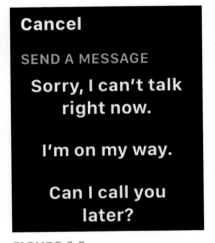

FIGURE 6-5

When you're in a conversation, the green answer button changes to the red end call button, as shown in **Figure 6-6**. Note the orange microphone icon next to the time in the upper-right corner of the same figure; it lets you know your Apple Watch's microphone is in use.

While on the call, you can do the following:

» Tap the mute button in the lower left to mute your Apple Watch microphone, as I did in **Figure 6-7** (the button turns white). Note that the orange microphone icon no longer appears next to the time in the upper right, indicating the microphone is not currently in use. Tap the mute button again to unmute the microphone and return to the call.

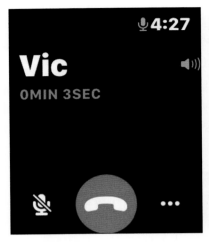

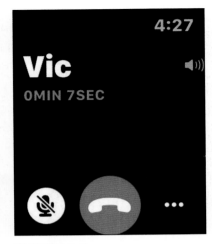

FIGURE 6-6

FIGURE 6-7

» Tap the more button (three white dots), and do one of the following:

- To open the keypad, tap the Keypad button (see **Figure 6-8**). This button comes in handy when you need to make selections in a menu while on a call.

- If you see more than one audio device listed, tap the audio device to use for your call. For example, in **Figure 6-9**, I have my AirPods Pro connected to my Apple Watch, so I can select either one to hold my conversation.

  Note that both options are available in Figure 6-9 because my call was muted. If my call had not been muted, the device currently being used for the conversation would have appeared dimmed.

» To adjust the volume of the audio, rotate the digital crown. As shown in **Figure 6-10**, the audio icon in the upper right turns green while you're adjusting the volume level.

» To hang up when you're finished with the conversation, tap the red end call button (refer to Figure 6-10).

TIP

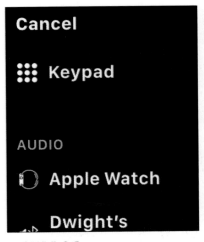

FIGURE 6-8

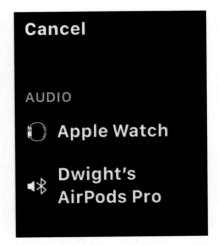

FIGURE 6-9

If you'd like to switch the call from your Apple Watch to your iPhone, unlock your iPhone and tap the green button in the upper-left corner for iPhones without Home buttons (see **Figure 6-11**) or tap the green bar at the top of the screen for iPhones with Home buttons.

Listen to voicemail

If a caller leaves a voicemail, you'll receive a notification on your Apple Watch. Swipe up on the display just a bit to see the playback controls for the voicemail, as shown in **Figure 6-12**.

You can do the following:

» To play the message, tap the play button (white arrow).

» To adjust the volume, rotate the digital crown.

» To return the call, tap the green call back button.

» To delete the voicemail, tap the red delete button.

» To send a message or dismiss the notification, scroll a bit farther down and tap the appropriate option (see **Figure 6-13**).

You can access voicemails also by opening the Phone app on your Apple Watch and tapping the Voicemail button.

TIP

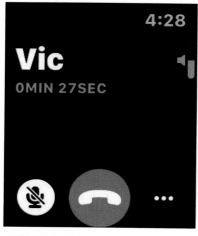

4:28

Vic

0MIN 27SEC

FIGURE 6-10

FIGURE 6-11

Make a call

Your Apple Watch can place calls as well as receive them.

REMEMBER

You can always use Siri to place calls, if you prefer. Simply say something like "Hey, Siri, call Mom" or "Hey, Siri, call 251-555-5555."

The Phone app on your Apple Watch will be where the action takes place. You'll find the Phone app button, shown in **Figure 6-14**, on the Apple Watch Home screen.

Let's get started making calls with the Apple Watch:

1. Open the Phone app on your Apple Watch by tapping its icon (refer to Figure 6-14).

FIGURE 6-12

FIGURE 6-13

2. Tap one of the calling options on the main Phone screen (shown in **Figure 6-15**):

- *Favorites:* When the Favorites list from your iPhone's Phone app appears, tap a favorite to call that person.

- *Recents:* Tap to call back one of your most recent callers.

- *Contacts:* Rotate the digital crown to scroll through the list of your iPhone's Contacts. Tap the one you'd like to call.

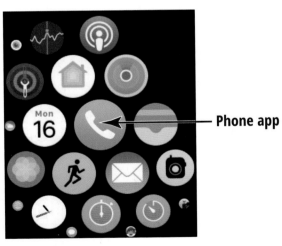

Phone app

FIGURE 6-14

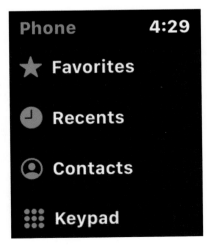

FIGURE 6-15

- *Keypad:* Enter a phone number by tapping the numbers on the digital keypad, as shown in **Figure 6-16**. Tap the green phone button in the lower-right corner to place the call, or tap the red delete button in the upper right to enter a different number.

Your Apple Watch will then initiate your call, like mine in **Figure 6-17**.

FIGURE 6-16

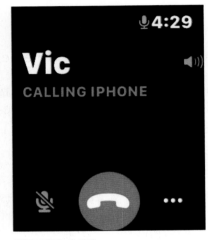

FIGURE 6-17

In addition to cellular and Wi-Fi calls, you can also make Face-Time audio calls with Apple Watch.

TIP

WHAT IF YOU HAVE TWO CELLULAR PLANS FOR YOUR iPhone?

Some iPhone models support dual SIM technology, which allows you to have more than one cellular plan on your iPhone — and allows you to use two cellular plans on your GPS + cellular Apple Watch too. When your Apple Watch is connected to your iPhone, you'll be able to use both plans with no problem, but when using it without your iPhone, you'll need to tell your Apple Watch which plan to use. For more information on this subject than I can cover here, visit Apple's Support site at https://support.apple.com/en-us/HT209043.

Emergency calls with Apple Watch

Apple Watch can place calls to your local emergency services (911, for example) with ease.

You can call emergency services from your Apple Watch in two ways: manually and automatically.

Place an emergency call

To manually place a call to emergency services:

1. Press and hold down the side button until the sliders shown in **Figure 6-18** appear on the display.

2. Slide the red Emergency SOS slider from left to right to place the call.

   After the call is over, your emergency contacts (if you have any) are notified and your location is sent to them. (Check out Chapter 10 for more info.)

FIGURE 6-18

To automatically call emergency services, press and hold down the side button until your Apple Watch starts beeping. A countdown also appears on the display. When the countdown reaches zero, your call will be placed.

WARNING

Don't use the automatic calling feature if you want to make your call in silence. The beeping will be heard, even if your Apple Watch is in silent mode.

Disable automatic calling

You can disable automatic calling. This may be a good idea if you have a habit of frequently pressing down the side button too long. Or if you keep your hands in your pockets a good bit, you might press the side button against the opening of your pocket and accidentally call emergency services (and, yes, I'm speaking from experience here).

On your Apple Watch, go to Settings ➪ SOS ➪ Hold Side Button, and then toggle the Hold Side Button switch, shown in **Figure 6-19**, to off (gray).

FIGURE 6-19

Get Your Message Across

Sending and receiving text messages using the Messages app on your Apple Watch is convenient and even fun on occasion. The convenience part comes with simply looking at your Watch when a message comes in and being able to tap one of the smart replies (prepared by the good folks in Cupertino) for a quick response. The fun part is when you reply with features like dictation and digital touch.

To get started, open the Messages app on your Apple Watch by tapping its icon (see **Figure 6-20**) on the Home screen.

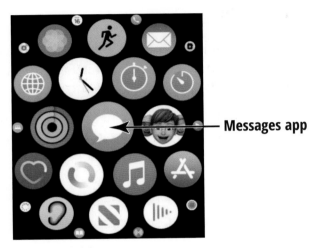

Messages app

FIGURE 6-20

Create a message

First, you learn how to create a message. Follow these steps:

1. Open the Messages app.

2. Tap the blue New Message button, shown in **Figure 6-21**.

3. Tap the blue + next to Add Contact, seen in **Figure 6-22**.

   You can add multiple contacts, if you like.

TIP

4. Add a contact to your message by tapping one of the three blue buttons:

   • *Dictation:* Say the name of the person (if the person is in Contacts) or say a phone number or an email address.

   • *Contacts:* Rotate the digital crown to scroll through your list of contacts.

   • *Keypad:* Enter a phone number using the keypad, and then tap the green OK button in the lower-right corner.

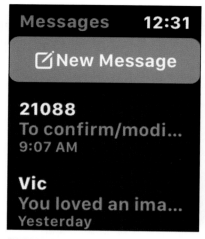

FIGURE 6-21

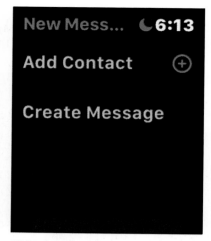

FIGURE 6-22

5. Tap the Create Message section.

6. Compose your message using one or more of the following tools:

- *Dictation:* Tap the microphone button, speak your message, and then tap the blue Send button in the upper right.

- *Scribble:* Tap the button that looks like a finger pointing to a letter, and drag your finger on the display in the shape of the first letter in your message. Each time you do so, if Apple Watch recognizes the character you've scribbled, it will add it to the message (with a cool ghostly animation, slightly shown in **Figure 6-23**) at the top of the display. When you've scribbled the entire message, tap the blue Send button in the upper right.

- *Emojis:* Tap the smiley face button, and then tap an emoji (see **Figure 6-24**) to send it along.

- *Memoji stickers:* Tap the memoji button (person's face with heart-shaped eyes), and then tap one of the memojis you've created on your Apple Watch or iPhone. You can then select from a variety of stickers that use your memoji, like I'm doing in **Figure 6-25**.

- *Digital touch:* Tap the button that looks like a heart with two fingers to send a sketch or an animation with haptics to convey your feelings. To send a sketch, just use your finger to draw on the display (as shown in **Figure 6-26**); tap the tiny button in the upper-right corner to change the color. You can send animations with haptics by using gestures to convey meanings.

FIGURE 6-23

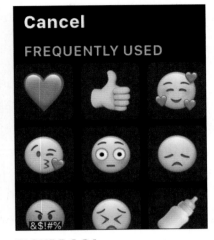

FIGURE 6-24

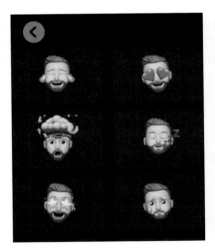

FIGURE 6-25

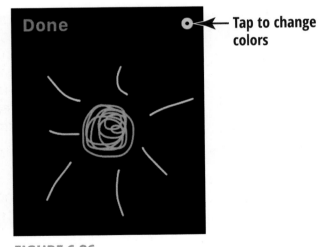

FIGURE 6-26

- *Smart replies:* Scroll down the display by swiping up or by rotating the digital crown to select from a list of smart replies that you can send without typing, dictating, sketching, or what-have-you.

7. Tap the blue Send button when you're ready to send your message along.

Read and reply to messages

You'll receive a notification on your Apple Watch when you get a message from someone. You can view the message right away by taping the notification, or you can view it later by tapping it in the Messages app.

To read and reply to messages:

1. Open the Messages app on your Apple Watch.

2. Tap the message to read or listen to it (if it's an audio message).

3. Scroll down by swiping up or by rotating the digital crown until you reach the end of the message.

   You'll see the buttons you need for your reply, as shown in **Figure 6-27**.

FIGURE 6-27

4. Tap one of the buttons to reply to the message (see the preceding section, "Create a message," for a detailed description of each), or tap one of the smart replies to use it.

   You can add your own smart replies! Open the Watch app on your iPhone, go to My Watch ⇨ Messages ⇨ Default Replies, and then tap Add Reply (at the bottom of the list). You can also edit and reorder the supplied replies; just click Edit in the upper right to start, and then tap Done in the upper right when finished.

5. Swipe or rotate the digital crown until you reach the bottom of the options to see the More section, as shown in **Figure 6-28**. From here, you can:

   - Tap Send Location to send a map of your current location. The recipient can tap it to open your location in Maps.

   - Tap Details to contact your friend: by phone, text message, email, or Walkie-Talkie (explained later in this chapter), as shown in **Figure 6-29**.

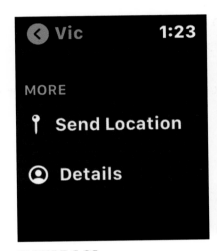

FIGURE 6-28

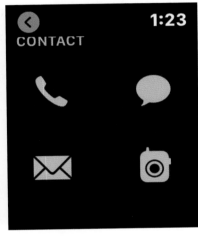

FIGURE 6-29

Yes, Virginia, You Can Email from Your Watch

As if phone calls and text messages from your Watch weren't cool enough, Apple makes sure you not only can read email but also reply and even compose emails from the handy-dandy device.

Compose an email

To compose a new email message:

1. Open the Mail app on the Home screen of your Apple Watch (see **Figure 6-30**).

2. Tap the blue New Message button, as shown in **Figure 6-31**.

FIGURE 6-30   FIGURE 6-31

3. Tap the Add Contact button.

You can add multiple contacts, if you like.

4. To add a contact to your email, tap one of the two blue buttons:

- *Dictation:* Say the name of the person (if the person is in Contacts) or the email address, and tap the Done button in the upper right to see a list of contacts that meet the description. Tap the one you want to use.

- *Contacts:* Rotate the digital crown to scroll through your contacts list.

5. Tap the Add Subject button.

6. Tap one or more of the following tools to add a subject line to your email:

- *Dictation:* Say the subject aloud then tap the Done button in the upper right (see **Figure 6-32**) to add the subject to your email's subject line.

- *Scribble:* Tap the button that looks like a finger pointing to a letter, and then drag your finger on the display in the shape of the first letter in your subject. If Apple Watch recognizes the character, it will add it to the subject line at the top of the display. When you've scribbled the entire subject, tap the blue Done button in the upper right.

- *Emojis:* Tap the smiley face button, and then tap one of the available emojis to use it as a subject.

7. Tap the Create Message section.

8. Compose your message using one or more of the following tools:

- *Dictation:* Tap this button, speak your message, and then tap the blue Done button in the upper right.

- *Scribble:* Tap this button, scribble the message, and then tap the blue Done button in the upper right.

- *Emojis:* Tap the emoji button, then tap one of the emojis to add it to your email message.

- *Smart replies:* Scroll down the display by swiping up or by rotating the digital crown to see a list of smart replies, and then add one to your message.

9. When you're ready to send your message along, tap the blue Send button (shown in **Figure 6-33**).

FIGURE 6-32

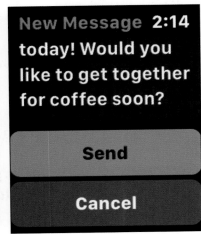

FIGURE 6-33

Read and reply to emails

Just as with text messages, you'll receive a notification on your Apple Watch when you get an email from someone. You can either view the notification now by tapping it or read it later by tapping it in the Mail app.

To read and reply to emails:

1. Open the Mail app on your Apple Watch.

2. Tap the email to read it.

3. Swipe up or rotate the digital crown until you reach the end of the message and see the options for working with the email, as shown in **Figure 6-34**.

FIGURE 6-34

These options are as follows:

- *Reply:* Tap to begin a reply to the sender.

- *Reply All:* Tap to begin a reply to everyone included on the original email.

- *Mark as Unread/Read:* Tap to mark the email as read or unread.

- *Archive Message:* Tap to archive the message.
- *Flag:* Tap to flag the message so you can mark it as important.

4. If you tapped Reply or Reply All, compose your reply message by using one or more of the following tools:

- *Dictation:* Tap this button, speak your message, and then tap the blue Done button in the upper right.
- *Scribble:* Tap this button, scribble the message, and then tap the blue Done button in the upper right.
- *Emojis:* Tap this button, and then tap an emoji to add it to your email message.
- *Smart Replies:* Scroll down the display by swiping up or by rotating the digital crown to see a list of smart replies, and then add one to your message.

5. When you're ready to send your reply, tap the blue Send button.

Put a Little Walkie in Your Talkie

The Walkie-Talkie app on your Apple Watch is another simple way to communicate with other Apple Watch users. It works just like a regular walkie-talkie and it is convenient to use when you just want to have a quick and easy conversation.

To use Walkie-Talkie:

1. Open the Walkie-Talkie app on your Apple Watch (the yellow icon shown in the center of **Figure 6-35**).
2. Tap a friend's name in your list.
3. Touch and hold down the talk button on your Apple Watch display while you talk, and then release it to let your friend talk back.

To invite friends to use Walkie-Talkie on their Apple Watch, tap the Add Friends button (shown in **Figure 6-36**), find them in your Contacts list, and tap to add them to Walkie-Talkie. After they accept the invitation, you're off and running.

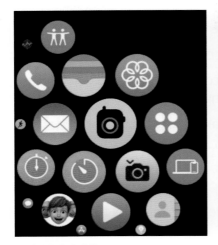

FIGURE 6-35

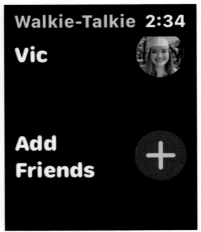

FIGURE 6-36

 TIP

You can turn Walkie-Talkie on or off by using the Walkie-Talkie toggle at the top of the menu in the Walkie-Talkie app.

Chapter **7**

Staying on Top of Things

L ife doesn't slow down, does it? I'm realizing that the older I get, the more stuff I have to do; it's just that today's stuff is different than yesterday's, or even last year's. Not only that, the more things I have to keep up with, the more tools I need to help me stay on top of them — the tools Apple Watch provides are the focus of this chapter.

You look at how Apple Watch can help you interact with your calendar, notifications, and reminders. You discover how to use several apps that can help you tackle daily activities. For example, Calendar and Reminders can help you stay on top of a lunch appointment, while Maps can guide you to the location of the restaurant. And if all else fails and you get turned around somehow, if the person you're meeting also has an Apple Watch, you can use the Find People app to pinpoint their location.

Keep On Schedule with Calendar

Calendar is an app that I can't live without. Whether it's the iOS version, macOS version, or in this case, watchOS version, this app is my lifeline. I rely on it to remember just about anything I have to do — I even have a Calendar event reminding me to add Calendar events (after typing that, I realize just how much I should see someone about this "problem"). Although not as full-featured as its counterparts, the watchOS version of Calendar is useful and handy in its own right. I rather enjoy escaping the tethers of my Mac and iPhone from time to time, and having Calendar on my wrist keeps me connected while not tying me to a larger, bulkier device.

You can work with Calendar events on your Apple Watch in several ways, so let's dive in.

View Calendar events

Viewing an event on your Apple Watch is super-simple:

1. Open the Calendar app on your Apple Watch. The default view for Calendar is called up next, as shown in **Figure 7-1**.

2. Use the digital crown to scroll backward or forward through your events.

3. Tap an event to see its details (see **Figure 7-2**).

4. Tap Today in the upper-left corner when you're ready to return to the up next view.

Calendar offers two other viewing options: List and Day. To change how your view events in Calendar:

1. Open the Settings app on your Apple Watch.

2. Tap Calendar, and then tap one of the view options:

   - *List,* shown in **Figure 7-3**, displays your upcoming events for the next few days in a convenient list. Use the digital crown or swipe up and down on the display to view the full list.

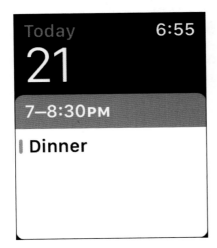

FIGURE 7-1

FIGURE 7-2

- *Day*, shown in **Figure 7-4**, displays your events for today in a time-line; use the digital crown or swipe up and down on the display to view the events for the full day. You can also swipe left or right to move from day to day.

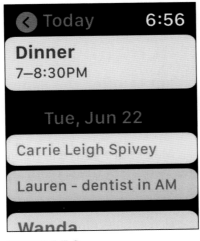

FIGURE 7-3

FIGURE 7-4

TIP

You can see the entire month when in list or day view by tapping the < in the upper-left corner. However, tapping a date on the month will not display events for that day, but will return you to today's events.

Add and delete events

Siri will help you add events to your Calendar, but you'll need to use a little elbow grease to delete them.

To add a Calendar event:

1. Simply say "Hey, Siri" (or press and hold down the digital crown) until Siri opens.

2. Say something along the lines of "Create a calendar event called Dinner with Cindy for tonight at 8:30 PM."

3. When Siri shows you the new event, tap the Confirm button to continue or the Cancel button to forget about it.

To delete an event:

1. Open the Calendar app on your Apple Watch.

2. Tap an event to open it.

3. Tap the Delete button at the bottom of the event details, and then tap Delete Event to confirm.

Customize Calendar event notifications

By default, your Apple Watch is set up to offer the same kinds of notifications for the same calendars as your iPhone, but you can customize those notifications for your watch:

1. Open the Watch app on your iPhone. Make sure you're in the My Watch tab; if not, tap My Watch at the bottom of the screen.

2. Tap Calendar.

3. Tap Custom in the Notifications section, the Calendars section, or both, and then select the types of notifications you want to see on your Apple Watch (as shown in **Figure 7-5**), as well as which calendars you want to view.

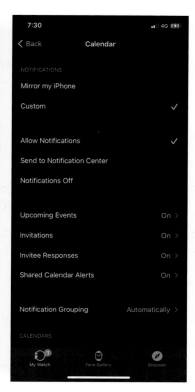

FIGURE 7-5

Notifications Aplenty

The title of this chapter is "Staying on Top of Things," and your Apple Watch uses notifications from apps to help you do just that. You can be notified of just about anything involving an app: messages, calls, new email, upcoming appointments — you name it. Apple Watch notifies you in a variety of ways, as well: sounds, haptic taps, and pop-up messages on your watch display.

Work with a notification

When you feel or hear a notification, just raise your wrist to see it (if it pops up while you're looking at your Apple Watch, that's all the more convenient). From here, you can scroll by swiping up and down on the notification to read all of it, or you can rotate the digital crown to do the same. Notifications usually allow you to pick at least one option (and sometimes many more), as shown in **Figure 7-6**. The one option that all notifications offer is the Dismiss button at the bottom.

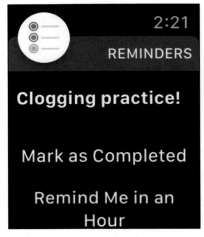

FIGURE 7-6

View notifications you've missed

Sometimes you don't want to view a notification or it sneaks through while you're doing something else. No worries — you can view those notifications in Notification Center. When you have unread notifications, a little red dot appears at the top of your Apple Watch display (see **Figure 7-7**).

Simply swipe down from the top of your display and Notification Center will open, allowing you to see notifications you've missed. You can swipe up and down the screen or use the digital crown to scroll up and down the list. Tap a notification to view it, like I'm doing in **Figure 7-8**; tap the Clear button to remove the notification. If you're viewing a group of notifications from a single app, swipe the notification you're viewing to the left and tap the X to clear it, or tap the Clear All button at the bottom of the group to clear all notifications in the group in one fell swoop.

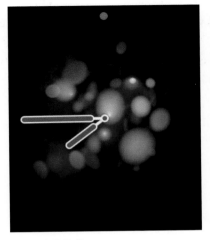

FIGURE 7-7

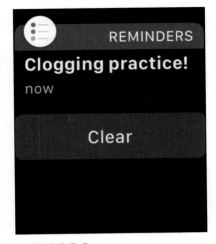

FIGURE 7-8

TIP

Need to be quiet, but you'd still like to know when notifications come in? Swipe up from the very bottom of your Apple Watch display to open Control Center, and then tap the silent mode icon (bell). Now when notifications arrive, you're notified only by haptic taps.

Customize notifications

Apple knows that while you may want to receive notifications one way on your iPhone, that doesn't necessarily mean you want to receive them in the same manner on your Apple Watch. That's why they offer you the ability to customize how your receive notifications for certain apps (Calendar, Mail, Messages, News, Phone, Photos, Podcasts, Reminders, Wallet and Apple Pay, and some third-party apps) on your watch. To customize notifications for those apps:

1. Open the Watch app on your iPhone. Make sure you're in the My Watch tab; if not, tap My Watch at the bottom of the screen.

2. Tap Notifications, and then tap the name of the app whose notifications you want to customize.

3. Tap Custom (Mirror My iPhone is the default), and then select one of the following:

- *Allow Notifications:* Notifications are displayed on your Apple Watch.

- *Send to Notification Center:* Notifications go straight to Notification Center and don't show up on the display or make a sound.

- *Notifications Off:* No notifications are sent from the app.

View and Create Reminders

Reminders is an app you'll find on every Apple device, and it helps you remember to get things done. From shopping lists to birthdays to dental appointments, Reminders will help you "stay on top of things" (there's that phrase again). You can create reminders on any of your Apple devices, and as long as you're using the same Apple ID on each one, the reminder will pop up on all of them when it's time, including your Apple Watch.

TIP

Mind you, the Reminders app is a bit sparse on the Apple Watch, but if you need to add or see more details in a reminder, just open Reminders on your iPhone, iPad, or Mac.

To view and work with reminders:

1. Open the Reminders app on your Apple Watch.
2. Swipe up or down, or rotate the digital crown, to view your reminder lists, as shown in **Figure 7-9**.
3. Tap a list to see the reminders it contains.
4. In the list you can:

- Mark a reminder as completed by tapping it; the circle to the left will fill in with the list's corresponding color.

- Tap the Add Reminder button at the bottom of the list to add a new reminder to the current list.

FIGURE 7-9

REMEMBER

Although you can't set dates and times for the reminder from here, you can do so using the Reminders app on your iPhone, iPad, or Mac.

TIP

You can also create reminders using Siri — and with Siri you are able to assign dates and times. Just say something like "Hey Siri, make a reminder to call the restaurant today at 1 PM."

- Tap View Options, and then either select Show Completed (to display all reminders for the list, even if you've marked them as Completed) or Hide Completed.

- Tap < in the upper left to return to the lists view.

REMEMBER

Any changes you make to your reminders on any Apple device will be reflected on your other devices, if they all use the same Apple ID. If you want to do tasks such as reordering your lists or deleting reminders completely, you'll need to do so on your iPhone, iPad, or macOS.

Voice Memos 101

The Voice Memos app is almost embarrassingly simple to use, but it does provide a valuable feature: the ability to record and play-back voice memos. I've used the Voice Memos app on my iPhone for years to help me remember things or to quickly record memorable moments, but have recently come to enjoy the convenience of using it on my Apple Watch.

TIP

Any voice memos you record with your Apple Watch will be available also on your iPhone, iPad, and Mac, as long as they're all signed in to the same Apple ID.

To easily record voice memos:

1. Open the Voice Memos app on your Apple Watch.

2. Tap the large red circle, which is unmistakably the record button.

3. When you've finished dictating, tap the stop recording button (see **Figure 7-10**).

Your new recording appears under the record button.

FIGURE 7-10

To easily play back voice memos:

1. Open the Voice Memos app on your Apple Watch.

2. Tap a recording to open it (they're found under the record button).

3. Tap the play button (see **Figure 7-11**) to listen to your prolific pronouncements, proclamations, and pontifications.

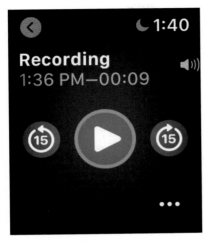

FIGURE 7-11

4. To return to your recordings list, tap < in the upper-left corner. If you instead want to send your recording to the scrap heap, tap the options icon (three dots) and then tap Delete.

You see now that I wasn't kidding when I described this app as simple but valuable.

Keep On Course with Compass

I do enjoy hiking, but I can't recall ever needing a compass to find my way around the trails. However, on the off-chance that I'll be lost in a South American rainforest, it's good to know that as long as I have my trusty Apple Watch with me (and as long as it maintains a charge), I'll be able to at least know north from south from east from west (and all points in between). "Why?" you ask? Because the watchOS team at Apple had the foresight to bequeath us Apple Watch users with a timeless tool: a compass.

I know I sound a little ho-hum about this one, but it's a cool app for what it does, if not an exceptionally simple one. It shows you not only which direction your Apple Watch is facing but also your current location and elevation.

TIP

Sorry, folks, but if your Apple Watch isn't a Series 5, Series 6, SE, or newer, you won't be able to take advantage of the Compass app.

To use Compass:

1. Open the Compass app on your Apple Watch.

2. To make sure your bearings are as accurate as possible, try to hold your Apple Watch as flat as possible to align the white crosshairs in the center of the compass.

   When aligned, the thin white lines of the crosshairs will appear as bold (or thick) white lines.

3. The red circle in the compass displays your elevation, incline, latitude, and longitude (as shown in **Figure 7-12**).

4. To view more information and to add a bearing, swipe up or down on the display, or rotate the digital crown.

FIGURE 7-12

TECHNICAL STUFF

To view coordinates and elevation, Location Services for your Apple Watch must be turned on (Settings ➪ Privacy ➪ Location Services) and Precise Location should be enabled for Compass on your iPhone (Settings ➪ Privacy ➪ Location Services ➪ Compass).

Check Your Math with Calculator

I remember thinking how absolutely cool the Casio Calculator Watch was when I was a wee lad. My morning routine always included a heated inner battle to determine whether I'd wear that or one of my brightly colored Swatch watches. It came down to who I was trying to impress: calculator watch for my fellow nerds, Swatch for the ladies.

Now that I'm all grown up, I still like to have a calculator on my watch, but this time it's 100 times cooler — I use the Calculator app on my Apple Watch. That's sure to impress everyone, right? Truthfully, it does come in handy when I need to figure out how much to tip in a restaurant and my foggy brain won't kick in with a quick answer. It's also saved my bacon when it comes to helping kids with homework. See, Casio had the right idea all along.

To use Calculator, just open the Calculator app on your Apple Watch and tap away.

Seriously, that's it. No need for further exposition or screenshots; it's a calculator.

Wait . . . did the Apple Watch come about as some sort of weird engineering experiment in Cupertino to see what would happen if you combined a Casio Databank, a Swatch, and an iPhone? Whoa, what an epiphany. But, I digress . . .

Stay Frosty with Weather

I love my Weather app. Our family lives on the Gulf Coast, and weather down here can be as fickle as a kid in an ice cream parlor. The Weather app on my Apple Watch keeps me up-to-date with just a glance.

You can always get the latest weather information quickly simply by asking Siri. Say something like, "Hey, Siri, what's the weather in Gulf Shores today?" Siri will then happily let you know the current conditions in Gulf Shores, as shown in **Figure 7-13**.

FIGURE 7-13

To see more info, swipe up or down or rotate the digital crown. There are nearly endless variations on how you can ask Siri about the weather conditions in your local area or around the world.

Get more weather information

If you prefer a more detailed report of the weather in your current location or a favorite city, you'll want to break out the Weather app:

1. Open the Weather app on your Apple Watch and you'll see a list of cities by default.

   If you're already viewing a city's information and want to return to the list, simply tap < in the upper-left corner.

2. Tap Viewing at the top of the list to determine whether your default view is set to temperature, conditions, or precipitation. Tap to choose one and return to the list.

3. Tap a city to view it's information. From here, you can:

   - View hourly forecasts based on temperature, condition, or precipitation, respectively, as shown in **Figure 7-14**. Tap the center of the clock to change the view.

   - To view more information, such as the air quality, UV index, or wind speed, swipe up and down or rotate the digital crown.

   - To find a 10-day forecast for the displayed city, swipe down or rotate the digital crown to the bottom of the information.

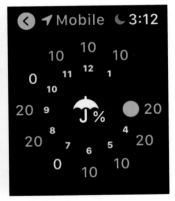

FIGURE 7-14

Add or remove cities

The default location for your Weather app is the same you've set for your iPhone. However, you can add or remove cities from your list directly from your Apple Watch (although one of the methods requires using your iPhone):

1. Open the Weather app on your Apple Watch.

   If you're viewing a city's information already, tap < in the upper-left corner to return to the cities list.

2. Scroll down or rotate the digital crown to the bottom of the cities list and then tap the Add City button.

3. Enter a city name by using dictation, scribbling it, or typing it using your iPhone's keyboard. If a list of multiple cities appears, tap to select the city you want to add.

   - *Dictation:* Tap the dictation icon (microphone), speak the name of the city, and tap Done.

   - *Scribble:* Tap the scribble icon (finger pointing to a letter), and drag your finger on the display in the shape of the first letter in the city's name. Each time you do so, if Apple Watch recognizes the character you've scribbled, it will add it to the city name (with a cool ghostly animation, slightly shown in **Figure 7-15**) at the top of the display. When you've scribble the entire name, tap Done.

FIGURE 7-15

- *Keyboard:* Tap the keyboard icon, and then use the keyboard on your iPhone to enter the name of the city. On your iPhone, tap the Apple Watch Keyboard notification that appears for opening the keyboard, and then type the name of your city and tap Search. The search results appear on your Apple Watch display.

4. To delete a city in your list, swipe it from right to left, and then tap the red X to send it on its way.

You can set a default city by going to Settings ➪ Weather ➪ Default City, and then tapping a city in your list.

On the Go with Maps

Luckily, I've always been good with directions and finding my way around unfamiliar places. However, for those rare times when I did get turned around, I've successfully used the Maps app to regain my bearings.

Maps allows you to do more than just see where you are — it can help you get where you're going. Maps can also find services in your neck of the woods, such as gas stations, coffee shops, and restaurants. You can also simply explore the area or get directions to a contact.

Explore your current location

If you want to find out where you are and what's around:

1. Open the Maps app on your Apple Watch.

2. Tap Location and a map of your current location appears, like the one in **Figure 7-16**. You're the blue dot in the center of the screen, and the blue fan indicates which direction on the map your Apple Watch is facing.

3. Move around the map:

- To move the map and view other items of interest in your area, simply drag the map with one finger.

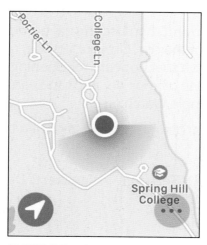

FIGURE 7-16

- Zoom the map in and out by rotating the digital crown. You can also double-tap the display, but the digital crown is a bit more elegant and accurate, at least to my taste.

- Tap a location or a landmark on the map to find out more about it, swiping or rotating the digital crown to view more info. Tap < in the upper-left corner to return to the map.

- To return to your current location, tap the location icon (white arrowhead in a blue circle; refer to Figure 7-16) in the lower left.

- To open the Search screen, tap the three dots (options button) in the lower-right corner, and then tap Search Here. More on that subject in the next section.

Search Maps

Maps allows you to search for locations and services right from your wrist.

To search for specific places or services by name:

1. Open the Maps app on your Apple Watch.

2. Tap Search, and then tap a button to enter information for your search:

- *Dictation:* Tap the dictation icon (microphone), speak the place or service you're looking for, and tap Done.

- *Scribble:* Tap the scribble icon (finger pointing to a letter), drag your finger on the display in the shape of the first letter of the place or service you're looking for. If Apple Watch recognizes the character you've scribbled, it will add it to the text field at the top of the display. After you've scribbled the words you need, tap Done.

- *Contacts:* Tap the contacts icon (blue circle with a white silhouette) and tap the name of a contact to view their information. Swipe or rotate the digital crown to find the address of the contact, and then tap the address. Select the mode of transportation you're using to get directions.

3. Tap an item in the search results to see information about it, such as a map to it (shown in **Figure 7-17**), directions based on the mode of transportation, and contact information.

FIGURE 7-17

To search for specific services by type:

1. Open the Maps app on your Apple Watch.

2. Tap Search, and then either swipe or rotate the digital crown to scroll to the list of services, as shown in **Figure 7-18**.

3. Tap an item in the search results to see information about it.

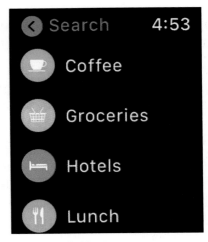

FIGURE 7-18

TIP

As always, Siri is at your disposal. Just ask Siri to find places or services near you and you will be rewarded (I hope) with a plethora of options.

Get directions

Getting directions is my favorite function of Maps, truth told. I love that the directions are almost always spot-on, and love even more that Siri can help by giving me turn-by-turn instructions, even tapping me on the wrist to alert me when it's time to turn or when I'm near my destination.

1. Open the Maps app on your Apple Watch.

2. Locate your destination in some fashion:

   - Tap Search, and then tap a button to enter information for your search (as described earlier in this section). Swipe or rotate the digital crown to the directions, and then tap a mode of transportation (such as walking, driving, or cycling).

   - Tap Search, tap a service from the list, and then tap the location you want directions to. Swipe or rotate the digital crown to the directions, and then tap a mode of transportation.

   - Ask Siri for directions to a specific place, contact, or address.

3. If offered more than one route, tap the route that suits you. View more information about a route by tapping the options button (three dots), which is shown in **Figure 7-19**. Swipe or rotate the digital crown to see information such as time and distance, as well as each step along the way. Tap Close in the upper-left to return to the suggested routes.

4. To begin your journey, tap the route you plan to use.

    Siri and Maps kick into action, displaying your route and letting you know when milestones along the trip are coming up, such as turns and when you've arrived.

5. If you want to get a look ahead, swipe or rotate the digital crown to a step in your directions, and then tap to see what that step looks like on the map, as shown in **Figure 7-20**. Tap < in the upper-left corner to return to the directions.

6. You can end directions before you arrive by tapping < in the upper-left corner of the directions, and then tapping the red End button at the bottom (you may need to swipe or rotate the digital crown to see it).

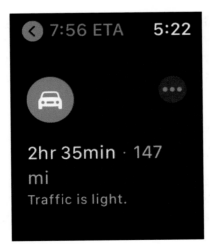

FIGURE 7-19

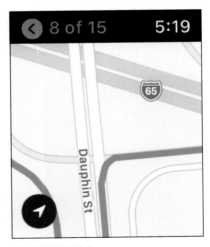

FIGURE 7-20

Find Folks

At first blush, the Find People feature may sound a bit creepy, but it's a helpful tool. Use it for meeting up with people in unfamiliar locations, keeping track of when someone leaves their location and arrives at another (super nice when you have kids in your life!), or letting someone else know when *you've* left or arrived somewhere.

Anyone with an Apple device can share their location with other Apple users, which is what allows this app to work its magic.

First, add a friend

The first step is to add a friend to your Find People app:

1. Open the Find People app on your Apple Watch.

2. Tap the Share My Location button, then tap either the dictation, contacts, or dial pad buttons to look up a contact or enter the phone number or email of the person you want to add.

3. Tap the appropriate option to choose how long to share your location with this person: for one hour, until the end of the day, or indefinitely, as shown in **Figure 7-21**.

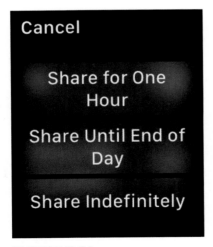

FIGURE 7-21

The person receives notification on their device that you want to share your location, and they can opt to share their location with you (how nice of them to reciprocate). At this point, both of you can use the Find People app to locate each other.

TIP
The equivalent to the Apple Watch's Find People app for iPhone, iPad, and Mac is called Find My.

Second, find a friend

Here's how to find folks using Find People:

1. Open the Find People app on your Apple Watch.

2. Tap the name of your friend in the list to see their location. Swipe or rotate the digital crown to get directions to their location or to send them a message.

3. Tap < in the upper-left corner to return to your list of friends.

Finally, get notified

You can also alert friends to when you leave your location or arrive at theirs, and vice versa:

1. Open the Find People app on your Apple Watch.

2. Tap the name of your friend in the list to see their location.

3. Tap Notify Me or Notify *friend's name.* Swipe or rotate the digital crown to see these options if you need to.

4. Toggle on (green) Notify Me or Notify *friend's name,* and then select whether to notify when someone has left or arrived, as shown in **Figure 7-22**.

TIP
If you decide you no longer want to share your location, open Find People, tap the friend you no longer want to share your location with, and then tap the Stop Sharing button.

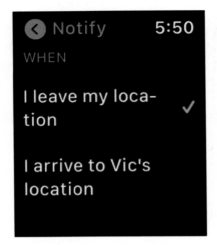

FIGURE 7-22

Chapter **8**

Meeting Siri

I'm a *Star Trek* fan from way back. Whether you duped it out from earlier comments I made in this book or another, or whether you're hearing this from me for the first time, it's a fact that I'm more of a "Space, the final frontier" kind of guy than a "Long ago in a galaxy far, far away" person. I certainly have enjoyed both sci-fi properties over the years, but the theremin from the original *Star Trek* theme still sings to me, if you will.

I remember watching Spock and Scotty talk to the computer, which spoke back — intelligently, no less! — carrying out their commands immediately and accurately. And the fact that I can now do something similar by talking to my watch or my phone is a bit trippy, in the best sense possible. Apple's introduction of Siri into their technological ecosystem about a decade ago is another of those marvels *Trek* is famous for having predicted, along with tablet computers (iPads, anyone?), universal translators (Apple's Translate app), teleconferencing (FaceTime), automatic doors, and more.

This chapter's aim is to introduce you to Siri and show you how you two can develop a terrific relationship using your Apple Watch to communicate.

Who Is This "Siri" You Speak Of?

Siri is a personal assistant that responds to the commands you speak to your Apple Watch, iPhone, iPad, Mac, and other Apple products. With Siri, you can ask for nearby coffee shops, and a list appears. You can dictate text messages rather than type them, which comes in handy with your Apple Watch. Ready to start a workout? Let Siri know what kind and it'll begin tracking it. You can open apps with a voice command. Calling your friends is as simple as saying, "Call Mateusz." Want to know the capital of New Zealand? Just ask. Siri checks several online sources to answer questions ranging from movie reviews to the next scheduled showing of the movie you're checking out at your local theater. Siri can keep you updated on the latest scores for your favorite team. Ask Siri to call family members and it'll have Memee or Popee on the line in no time flat. You can also have Siri perform other tasks as well, such as returning calls and controlling music and other multimedia.

watchOS 8 introduces several updates that allow Siri to do more for you and to do it faster than ever before. Dictation and commands used to be passed from your Apple Watch to your iPhone and then back again, but now a bulk of that work is done directly on your Apple Watch, increasing accuracy and speed. Siri can even translate into ten languages, right from your wrist.

"Beam me up!" (I can't wait until Siri can execute *that* command!)

Set Up Siri for Apple Watch

As I've stated, you must have an iPhone if you want to use an Apple Watch, and Siri is one reason why. You must have Siri activated on your iPhone for it to work with your Apple Watch.

TIP

Siri is available only if your Apple Watch has internet access via your iPhone or Wi-Fi, or if your Apple Watch is a GPS + Cellular model. Cellular data charges could apply when Siri checks online sources if you're not connected to Wi-Fi. In addition, available features may vary by area.

If you didn't activate Siri during your iPhone's registration process, you can use Settings to turn Siri on by following these steps:

1. Tap the gear icon (Settings) on the Home screen.

2. Tap Siri & Search (see **Figure 8-1**).

3. In the Siri & Search screen in **Figure 8-2**, toggle the on/off switch to on (green) to activate any or all of the following features:

   - If you want to be able to activate Siri for hands-free use, toggle the Listen for "Hey Siri" switch to turn on the feature. When you first enable Hey Siri, you'll be prompted to Enable Siri. Just tap the Enable Siri button, walk through the steps to enable it, and continue.

     With this feature enabled, just say "Hey, Siri" and Siri opens, ready for a command. In addition, with streaming voice recognition, Siri displays in text what it's hearing as you speak, so you can verify that it has understood you correctly.

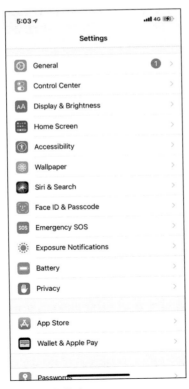

FIGURE 8-1

FIGURE 8-2

- Press Home for Siri requires you to press the Home button to activate Siri. Alternatively, if your iPhone doesn't have a Home button, you'll see Press Side Button for Siri, requiring the press of the side button to activate Siri.

- Allow Siri When Locked allows you to use Siri even when the iPhone is locked.

4. If you want to change the language Siri uses, tap Language and choose a different language in the list that appears.

    This setting will affect both your iPhone and Apple Watch initially. You can change the language on your Apple Watch later, if you prefer.

5. To change the nationality or gender of Siri's voice from American to British or Australian (for example), or from female to male, tap Siri Voice and make your selections.

    This setting will also initially set the defaults for both your iPhone and Apple Watch, but again, you can change the Apple Watch settings later, if you want.

6. Let Siri know about your contact information by tapping My Information and selecting yourself from your contacts.

Now that you have Siri set up on your iPhone, let's take a look at it on your Apple Watch. Siri should be automatically enabled by default:

1. Open the Settings app on your Apple Watch.

2. Tap Siri (see **Figure 8-3**).

3. In the screen in **Figure 8-4**, toggle the on/off switch to on (green) to activate any or all of the following features:

    - If you want to be able to activate Siri for hands-free use, toggle the Listen for "Hey Siri" switch to turn on the feature. If Siri is disabled on your Apple Watch, or when you first enable Hey Siri, you'll be prompted to Enable Siri. Just tap the Enable Siri button to continue.

        With this feature enabled, say "Hey, Siri" and (just like on your iPhone) Siri will open up and await your command. As with your iPhone, with streaming voice recognition, Siri displays in text what it's hearing as you speak.

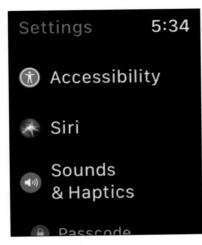

FIGURE 8-3

FIGURE 8-4

- Raise to Speak allows you to activate Siri simply by raising your arm. To give Siri a command, hold the watch close to your mouth and speak; there's no need to even say "Hey, Siri" when using this feature. Toggle the switch on to enable Raise to Speak.

- Toggle on the Press Digital Crown option, which allows you to activate Siri by pressing and holding down the digital crown.

4. If you want to change the language Siri uses on your Apple Watch (this won't change the language on your iPhone), tap Language and choose a different language in the list that appears.

5. To change the nationality or gender of Siri's voice for your Apple Watch, tap Siri Voice and make your selections.

    This setting affects Siri's nationality or gender or both only on your Apple Watch.

6. Let Siri know how it should reply to you with its voice by using the options in the Voice Feedback section. Just tap one of the following:

   - *Always On:* Siri is able to speak to you regardless of whether or not the Apple Watch is in silent mode.

   - *Control with Silent Mode:* Siri will speak replies only when your Apple Watch is not in silent mode.

   - *Headphones Only:* Siri will speak replies only when you have headphones connected to your Apple Watch, so that only you hear them.

TIP

You're still able to read Siri's replies on your Apple Watch display, even when you've elected not to hear voice replies.

Understand What Siri Can Do on Apple Watch

Siri on Apple Watch isn't as full-featured as it is on iPhone, but it is still powerful and useful. For example, you can't send email from Siri by using your Apple Watch, but you can from your iPhone.

You can pose questions or ask Siri to do something such as make a call or add an appointment to your calendar. Siri can also search the internet or use an informational service called Wolfram Alpha to provide information on just about any topic.

Siri requires no preset structure for your questions; you can phrase things in several ways. For example, you might say, "Where am I?" to see a map of your current location. Or you could say, "What is my current location?" or "What address is this?" and get the same results.

Siri also checks with Wikipedia, Bing, and Twitter to get you the information you ask for. In addition, you can use Siri to tell Apple Watch to control playback in the Music app, dictate text messages, and much more.

Siri knows what app you're using, though you don't have to have that app open to make a request involving it. However, if you are in the Messages app, you can make a statement like, "Tell Victoria I'll be late," and Siri knows that you want to send a message.

If you ask a question, Siri responds to you both verbally and with text information (see **Figure 8-5**), or by opening a form, as with texts, or by providing a graphic display for some items, such as maps. When a result appears, you can tap it to make a choice or open a related app.

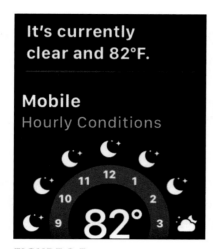

FIGURE 8-5

Siri works with most apps on your Apple Watch, including Music, Messages, Calendar, Maps, Weather, Settings, and Podcasts. In the following tasks, I provide a quick guide to some of (but by no means all!) the useful ways you can utilize Siri on your Apple Watch.

TIP

Siri for watchOS 8 now supports many different languages for translation, so you can finally show off those language lessons you took in high school. Siri can translate English into Mandarin, French, German, Italian, Spanish, Arabic, Russian, Portuguese, Japanese, and Korean, all from your Apple Watch!

REMEMBER

No matter what kind of action you want to perform, you must activate Siri before speaking your command or question. Raise your arm and place your watch near your mouth to use Raise to Speak; press and hold down the Home button (or side button for iPhone models without a Home button) until Siri opens; or if you've enabled Hey Siri, simply say the phrase.

Call contacts

First, make sure that the people you want to call are entered in the Contacts app on your iPhone and that their phone numbers are in their record. Then, follow these steps:

1. Activate Siri on your Apple Watch using whatever method you prefer.

2. Speak a command, such as "Call Reid" or "Call Mom."

3. If you have two contacts who might match a spoken name, or a contact with multiple phone numbers, Siri responds with a list of possible matches. Tap one in the list or state the correct match to proceed.

   The call is placed.

4. To end the call before it completes, tap the end call button (white phone in a red circle, as shown in **Figure 8-6**).

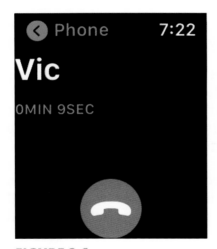

FIGURE 8-6

To cancel any spoken request, say "Cancel" or press the digital crown once.

Create reminders and alerts

You can also use Siri to create reminders and alerts:

1. Activate Siri.

2. Speak a command, such as "Remind me to call Dad on Thursday at 10 a.m." or "Wake me up tomorrow at 6:15 a.m."

   A preview of the reminder or alert is displayed (see **Figure 8-7**).

3. If you change your mind, tell Siri "Cancel" or "Remove."

FIGURE 8-7

Add tasks to your calendar

You can use Siri also to set up events on your calendar:

1. Activate Siri.

2. Speak a phrase, such as "Set up a meeting for 3 p.m. tomorrow."

3. Tap Confirm to add it to your calendar. Siri sets up the appointment (see **Figure 8-8**). Tap Cancel to forget the whole affair.

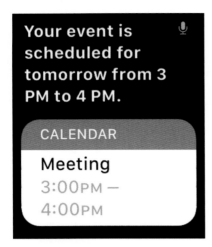

FIGURE 8-8

Play music

You can use Siri to play music from the Music app on your Apple Watch or iPhone:

1. Activate Siri (are you seeing a pattern here?).

2. To play music, speak a command such as "Play music" or "Play Jazz radio station."

   You can also ask Siri to play a specific song, album, or radio station; Siri responds, as shown in **Figure 8-9**.

If you're listening to music or a podcast with Bluetooth head-phones connected and then stop midstream, the next time you reconnect the headphones, Siri starts playing where it left off.

Get directions

You can use the Maps app and Siri to find your current location, get directions, find nearby businesses (such as restaurants or banks), or get a map of another location. Be sure Location Services is enabled on your Apple Watch, to allow Siri to know your current location. (Go to Settings ⇨ Privacy ⇨ Location Services, and be sure the toggle switch is set to on.)

FIGURE 8-9

TIP

Also be sure to allow Location Services with Siri & Dictation in your iPhone's Settings app, too. Go to Settings ⇨ Privacy ⇨ Location Services ⇨ Siri & Dictation, and then tap While Using the App.

Here are some of the commands you can try to get directions or a list of nearby businesses:

» **"Where am I?":** Displays a map of your current location.

» **"Where is Washington Square?":** Displays options for directions based on your mode of travel, as well as a map of that park's location (see **Figure 8-10**).

» **"Find pizza restaurants.":** Displays a list of restaurants near your current location; tap one to see directions, a map of its location, and more info.

» **"Find PNC Bank.":** Displays a list of locations for the indicated business (or in some cases, several nearby locations, such as a bank branch and all ATMs). Tap one to see directions and a map.

» **"Get directions to the Roman Colosseum.":** Provides directions to the site from your current location.

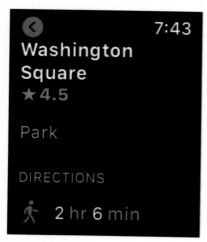

FIGURE 8-10

Ask for facts

Wolfram Alpha is a self-professed online computational knowledge engine. It's more than a search engine because it provides specific information about a search term rather than multiple search results. If you want facts without having to spend time browsing websites to find those facts, Wolfram Alpha is a good resource.

Siri uses Wolfram Alpha and such sources as Wikipedia and Bing to look up facts in response to questions such as "What is the capital of Kansas?" "What is the square root of 5,000?" or "How large is Pluto?" Just activate Siri and ask your question; Siri consults its resources and returns a set of relevant facts.

You can also get information about other things, such as the weather, stocks, or the time. Just activate Siri and say a phrase like one of these to get what you need:

» **"What is the weather?":** Get the weather report for your current location. If you want weather in another location, just specify the location in your question.

» **"What is the price of Apple stock?":** Siri tells you the current price of the stock or the price of the stock when the stock market last closed.

» **"What's the temperature of the sun?":** Siri tells you the temperature of the sun (see **Figure 8-11**), and provides much more, such as the sun's density, weight, radius, and distance from the Earth.

FIGURE 8-11

Send messages

You can create text messages by using Siri and existing contacts:

1. Activate Siri.

2. Speak a command to Siri. For example, if you say, "Text Emi" or "Send a message to Devyn," Siri asks what you want to say.

3. Speak your message contents and Siri displays it for you on your Apple Watch.

   If it looks good, leave it alone and Siri will send it on its way. If you don't want to send it, tap the Don't Send button (you may need to swipe or scroll down a bit to see the button).

TIP

You can't send digital touch or animated emoji messages by using Siri, but you can with the Messages app itself.

The fun with Siri doesn't end there. Keep trying out Siri with all kinds of tasks and various questions, and you may be surprised how quickly you integrate it into your everyday life.

Chapter **9**

It's a Wallet, Too?

Wouldn't it be super-cool and super-convenient if you could send cash to someone with just a few taps on a screen? Or how about if you could pay for gasoline at the pump, even if you left your wallet at home? Can you imagine being able to buy movie tickets, choose your seats, and have those tickets stored on your phone, while watching your grandkids' soccer game?

Wouldn't it be great to go shopping and not have to pull out a credit or debit card everywhere you went? Oh, wow, it would be even better if you could simply hold your watch up to a device and pay instantly, without digging through your purse or wallet for those very cards! I think we're on to something.

Now all we need is for some global super-company with a bevy of insanely great engineers to put this all together and make it a new way of life for all who dare move even further into the brave new connected world of the 21st century. You know, Apple CEO Tim Cook is from a small town just across the bay from where I sit typing this. Maybe I should have my people give his people a call?

Uh-oh, it seems our great idea is already a thing! Apple Wallet, Apple Pay, and Apple Watch (and let's throw the iPhone in the mix, too) can already do what I just described and more. In this chapter, I show you how to take advantage of these exciting developments!

Discover Apple Wallet and Apple Pay

Apple Wallet and Apple Pay work hand-in-hand in a harmonious relationship to make your life easier. Apple Wallet carries your stuff, while the Apple Pay system securely allows you to add items such as credit cards, debit cards, and store cards to your Wallet.

Apple Wallet overview

Apple Wallet is simply an app for storing digital versions of payment methods and other items that you might carry in a real wallet. You need to use the Apple Wallet app on your iPhone to get cards and other documents into the Apple Watch, but after that you can use your Watch independently of your iPhone.

The Wallet app, shown in **Figure 9-1**, hangs out on the Apple Watch Home screen.

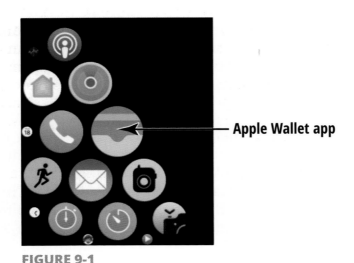

Apple Wallet app

FIGURE 9-1

TIP

Keep your iPhone handy while reading this chapter if you want to follow along with the instructions.

Apple Wallet supports these cards and passes:

» Credit and debit cards

» Prepaid cards

» Store and rewards cards

» Transit cards

» Coupons

» Boarding passes

» Driver's licenses

» Insurance cards

» Event tickets (movies, concerts, and so on)

» And much more

Apple Pay overview

Apple Pay is a secure and private way to make purchases and exchange other information. It's the technology that helps you add cards and passes to Apple Wallet and then safely use them — typically in a contactless manner — to pay for things in stores or online, gain access to events, and just about anything else for which you'd typically use a card for. As of this writing, Apple claims that Apple Pay is accepted at more than 85 percent of the retailers in the United States, which means you can use it just about anywhere.

TIP

To find out which banking institutions support their cards on Apple Pay, visit https://support.apple.com/en-us/HT204916.

JUST HOW SECURE IS APPLE PAY?

You might be wondering whether using Apple Pay for financial transactions is safe and secure. According to Apple, it's better to use Apple Pay than cash or physical cards because the transactions are private. Apple doesn't keep track of the card numbers or your transaction information. That info is between you, the merchant, and your bank. Please don't just take my word for it, though: Check out `https://support.apple.com/en-us/HT203027` for Apple's detailed explanation on how they handle your information when you're using Apple Pay.

To give you an idea of how simple it is to use Apple Pay, here's a quick overview of how it works with your Watch:

1. Use the Apple Wallet app on your iPhone to add a card or pass to it.

   A copy of the card or pass is stored on your Apple Watch, too.

2. When ready to use the card or pass, simply open Apple Wallet on your Apple Watch and hold it up to the contactless reader.

   The process is complete when you feel a tap on your wrist.

Now you can simply repeat Step 2 wherever you go.

REMEMBER

Apple doesn't charge a fee to use Apple Pay — it's a free service.

Set Up Apple Wallet

Without cards, Apple Wallet is no more useful than an empty physical wallet. In the first part of this section, you find out how to add cards to Wallet, as well as how to set up Apple Wallet for your Apple Watch. The last part of this section discusses adding passes and other types of documentation.

Add cards to Wallet

Adding cards to Wallet is a breeze, but you'll need to break out your iPhone to do so.

To add new cards to Wallet:

1. Gather the card(s) you'd like to add to Wallet.

2. Open the Apple Wallet app on your iPhone by tapping its icon on the Home screen (see **Figure 9-2**).

Apple Wallet app

FIGURE 9-2

3. To begin adding a new card, tap the + button in the upper-right corner of the app (see **Figure 9-3**).

4. Select the type of card you want to add (see **Figure 9-4**):

- Tap Debit or Credit Card to add one of these.

- Tap Transit Card to add a supported transit card for your region. Learn more about transit cards and Apple Wallet by visiting `https://support.apple.com/en-us/HT207958`.

Tap to add a card

FIGURE 9-3

FIGURE 9-4

5. Scan your card or manually enter its information.

6. Follow the on-screen instructions, which may differ depending on the card type.

TIP

Some banks require a bit more work than others to add their cards to Apple Wallet, so don't be surprised if you have to take additional steps for some than you do for others.

Set up Apple Wallet for Apple Watch

Now it's time to get your cards into your Apple Watch. To do so:

1. Open the Apple Watch app on your iPhone.

2. Tap the My Watch tab at the bottom of the screen.

3. Find and tap Wallet & Apple Pay, shown in **Figure 9-5**.

   The screen shown in **Figure 9-6** appears, and the Payment Cards on Your Watch section lists cards already on your Apple Watch, if any.

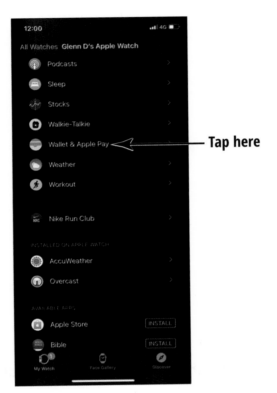

FIGURE 9-5

FIGURE 9-6

4. If you see cards that you've added on other devices and need to add to this iPhone, just tap the Add button that appears next to them, and then enter the CVV code from the back of that card when prompted.

5. To add a new card, tap the orange Add Card button, select the type of card you want to add, scan the card or manually enter its information, and then follow the on-screen instructions.

6. To make a card your default when using your Apple Watch:

   a. In the Transaction Defaults section, tap Default Card. The screen shown in **Figure 9-7** appears.

   b. Tap the card you want to use as your default, and then tap Back in the upper-left corner.

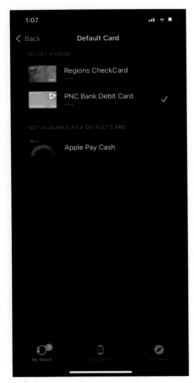

FIGURE 9-7

REMEMBER

You can still use other cards in your Apple Wallet for Apple Watch. Setting one as the default only means it's the first one to pop up when you go to pay for something.

Use Apple Watch for purchases

Now that you've added cards, it's time to starting spending that money! (Wisely, of course.)

To perform a transaction using Apple Wallet on your Apple Watch:

1. Double-click the side button on your Apple Watch or open the Apple Wallet app on your Watch.

   Your default card will appear, as shown in **Figure 9-8**.

2. If you don't want to use the default card, rotate the digital crown or swipe up or down on the Apple Watch display until you see the one you want.

3. Hold your Apple Watch near the contactless reader.

   When the reader and Watch conduct the transaction successfully, you will hear a beep, feel a tap, and get a notification in Notification Center.

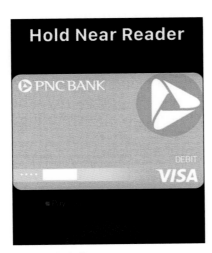

FIGURE 9-8

TIP

To view information about a card on your Apple Watch or to delete it from your Watch, tap the card on your display, and then swipe up and tap Card Information or Delete.

TIP

You can also make a purchase in an app on your Apple Watch. When it comes time to choose a payment method, just tap Apple Pay and then double-click the side button when prompted.

WARNING

If you lose or misplace your Apple Watch, don't mess around! If you can't find it quickly, you should put your Apple Watch in lost mode, which will lock it down and prevent others from using it to make purchases. You can also take other steps; please visit `https://support.apple.com/en-us/HT207024` at Apple's Support site for much more information.

Add passes and other items to Wallet

I absolutely love that I can add items to Wallet other than just financial cards. I have my driver's license, auto insurance card, rewards cards (such as my Delta SkyMiles card), store cards (like Starbucks), and others stored. I also use Apple Wallet for event tickets, boarding passes, and other cool things. It's so convenient to be able to walk up and scan my Watch display, as opposed to fumbling and bumbling with paper tickets and documents!

Links or instructions to add passes and other items to Wallet generally come from the organization that you're using them for. For example, when I buy movie tickets using the AMC app on my iPhone, I see a button to add the tickets to Apple Wallet, as shown in **Figure 9-9**.

Other ways an organization might send a link are

» By email

» Through an app

» With a notification

» Using Messages

To use a pass or other item that contains a barcode:

1. Open the Apple Wallet app on your Apple Watch, or double-click the side button.

Tap to add to Wallet

FIGURE 9-9

2. Scroll through your cards and passes until you get to the item you want to use, as shown in **Figure 9-10**.

3. Tap the card or item to reveal the barcode.

   If you need to exit the barcode to go back to your pass, just scroll down using the digital crown to reveal the pass again, like I'm doing in **Figure 9-11**.

4. Hold the barcode up to the contactless scanner. Done!

TIP

For a more in-depth discussion of everything you can do with Apple Pay, please check out Apple's Support site at https:// support.apple.com/en-us/HT201239. This page can show you how to take full advantage of Apple Pay's awesomeness and can go much further than the confines of this little book.

FIGURE 9-10

FIGURE 9-11

3

Just for the Health of It

IN THIS PART . . .

Setting up the Health app

Discovering apps to monitor your heart and breathing

Keeping tabs on sleep and noise levels

Tracking your daily activity

Working out with Apple Fitness+

Chapter **10**

Tracking Your Health

As we get older, the need to keep track of just how we're doing physically becomes increasingly more apparent. Maybe it's not quite as easy as it once was to navigate that flight of stairs in your home, or perhaps chasing your dog around the park is a bit more taxing than you remember it being in months past. But getting older doesn't mean you're sentenced to spend your remaining days in a recliner watching reruns on TV-land (not that there's anything wrong with that from time to time). It just means that we need to be more vigilant not only with exercise and proper diet but also with the signals the body gives us that something isn't quite right.

To my mind, one of the most natural functions of the Apple Watch is to assist with health monitoring and activities. So much of our health can be tracked by this little miracle because the wrist on which it resides is a hub for a wealth of information about our body's current state of affairs. From the first time Apple mentioned the idea of the wearable device, it was crystal clear that health was and would continue to be a major point of emphasis. In the years since, Apple has backed up that assertion by providing more health-related apps and hardware, and by teaming up with healthcare providers and fitness partners to make the Apple Watch an integral tool in the healthcare arsenal of millions.

In this chapter, you look at how Apple Watch can help you keep tabs on your health, and get to know the health-related apps that come standard with watchOS and iOS for your iPhone.

TIP

Because your iPhone and Apple Watch work together for so much of these health-related functions, I discuss the iPhone a bit more in this chapter than others. You'll want to have it handy while reading.

WARNING

As great as the Apple Watch is at helping you keep track of your health, it's no substitute for your doctor. If you're not feeling well or simply think something could be wonky with your health, please don't rely on your Apple Watch health data; seek the advice of a medical professional.

Set Up a Medical ID

Sometimes in a medical emergency, you may not be able to provide critical information to caregivers; that's where your medical ID comes in. Your medical ID lists important facts about you and your medical history on your Apple Watch so caregivers can know how to best help you. For example, if you have allergies or a history of medical issues, putting that info in your medical ID could be crucial in an emergency situation.

To set up your medical ID, you'll need your iPhone:

1. Open the Health app on your iPhone and make sure you're in the Summary tab. (If not, tap it at the bottom of the screen.)

2. Tap the profile picture in the top right, and then tap Medical ID.

3. Tap the blue Get Started button, and then begin filling in the information as indicated (as seen in **Figure 10-1**).

4. When you've finished entering information, tap Done in the upper-right corner.

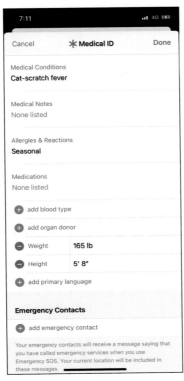

FIGURE 10-1

Please try to enter as much information as possible. Also, I want to emphasize the following two items in your medical ID:

>> **Emergency Contacts:** When you use Emergency SOS to place an emergency call (see Chapter 6 for more on that), folks you list here will also be alerted and your location will be provided to them (if it's available).

>> **Emergency Access:** This section has two options to consider:

- The Share During Emergency Call option allows your iPhone and Apple Watch to share your Medical ID information with emergency services (such as 911). If this option sounds good to you, toggle the switch on (green).

- The Show When Locked option is non-negotiable in my opinion, because it allows medical personnel to view your medical ID info even when your iPhone is locked. They can also see your medical ID by holding down the side button on your Apple Watch until the sliders appear, and then sliding the Medical ID slider (shown in **Figure 10-2**) from left to right.

FIGURE 10-2

TIP

If the Medical ID slider doesn't appear when you hold down the side button, that option is disabled in the Watch app on your iPhone. To enable it, open the Watch app, go to My Watch ⇨ Health ⇨ Medical ID, tap Edit, toggle on (green) the Show When Locked switch, and then tap Done.

Check Out Your iPhone's Health App

The Health app is the one-stop shop for your health information. The app can collect data from other health-focused apps and from equipment that works with the Health app, and you can also input data manually.

The Health app is divided into three parts: the Summary tab, the Sharing tab, and the Browse tab.

Summary tab

The first thing you'll want to do with the Health app is set up your health profile, shown in **Figure 10-3**. Your health profile contains the most basic information about you, such as your name, date of birth, gender, height, and weight.

FIGURE 10-3

To set up your health profile:

1. Open the Health app on your iPhone and make sure you're in the Summary tab. (If not, tap it at the bottom of the screen.)

2. Tap the profile picture in the top right, and then tap Health Details.

3. Tap the blue Edit button in the upper-right, and then tap a subject to begin filling in your pertinent information.

4. When you're finished, tap Done in the upper-right corner, tap Profile in the upper left, and then tap Done in the upper-right to return to the Summary screen.

The Summary screen, shown in **Figure 10-4**, is an overview of your health-related data, and may also include messages and alerts.

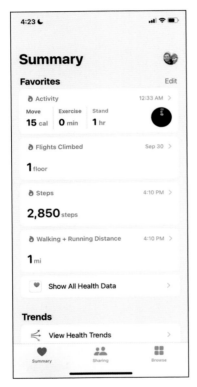

FIGURE 10-4

The Favorites area lists data collected about your health with other apps and devices. Tap any of the entries in the Favorites area to see more detail of the item or activity.

You can customize what shows up in the Favorites area:

1. Tap the blue Edit button to the right of the word Favorites.

2. Tap the Existing Data tab or the All tab at the top.

Existing Data shows only options for data you've accumulated thus far. All lists all possible data options that Health can display.

3. Tap the star to the right of data you want to add (the star will become solid blue) or remove (the star will turn to white), as shown in **Figure 10-5**.

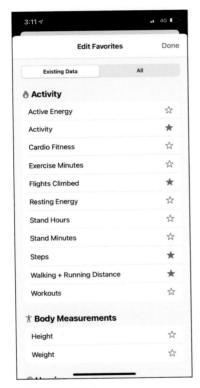

FIGURE 10-5

4. When you're finished, tap the Done button in the upper right.

The Favorites section displays the content you've requested.

Scroll through the Summary page to find other areas, such as Highlights, Get More from Health, and Apps. Highlights shows you achievements you've made recently, such as walking more steps today than yesterday. Get More from Health alerts you to other ways you can maximize the Health app, such as by setting up the Cardio Fitness Levels or Noise Notifications health-related task on your

Apple Watch. The Apps area, shown in **Figure 10-6**, lists featured third-party apps that utilize HealthKit, which makes them compatible with the Health app and other apps and devices that use HealthKit. Just tap the name of an app in the list to see more information and to install it on your iPhone and Apple Watch (if a version is available for Apple Watch).

FIGURE 10-6

Sharing tab

The iOS 15 iteration of the Health app allows you to share your health data with anyone in your contacts and even directly with your doctor. This is a great idea if you'd like to keep family and friends abreast of any changes you may experience. Best of all, you're in complete control of what you share and when to stop sharing. Any data you share with others is also encrypted, meaning that it's shielded to protect you from internet ne'er-do-wells. To begin sharing with a contact:

1. Tap the Sharing tab, then tap the Share with Someone button.

2. Enter the contact's name in the Search field, then tap the name of the contact.

   If their device supports sharing, it will be listed in blue; if not, it will be listed in gray.

3. Tap either the See Suggested Topics button (to see a list of topics to share suggested by the Health app) or the Set Up Manually button (to manually select topics you want to share).

4. Toggle the switches on for the topics you want to share.

5. After you've selected the shared topics, tap Share and then tap Done.

REMEMBER

You can share with other contacts only if they're running at least iOS 15 on their iPhone.

For more information on sharing your health data, including how to share this info with your doctor, visit `https://support.apple.com/en-us/HT212629`.

Browse tab

The Health app organizes your data by categories: Activity, Body Measurements, Cycle Tracking, Hearing, Heart, Mindfulness, Mobility, Nutrition, Respiratory, Sleep, Symptoms, Vitals, and Other Data (such as inhaler usage and handwashing). Tap the Browse tab at the bottom of the screen to see a list of these categories (see **Figure 10-7**).

These categories may also list subcategories, which you can see by tapping a main category, such as Mobility (shown in **Figure 10-8**).

TIP

Learn more about the Health app by visiting Apple's Health website: `www.apple.com/ios/health`.

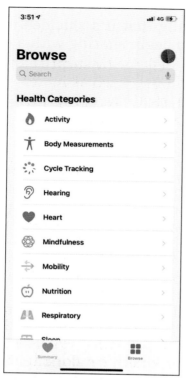

FIGURE 10-7

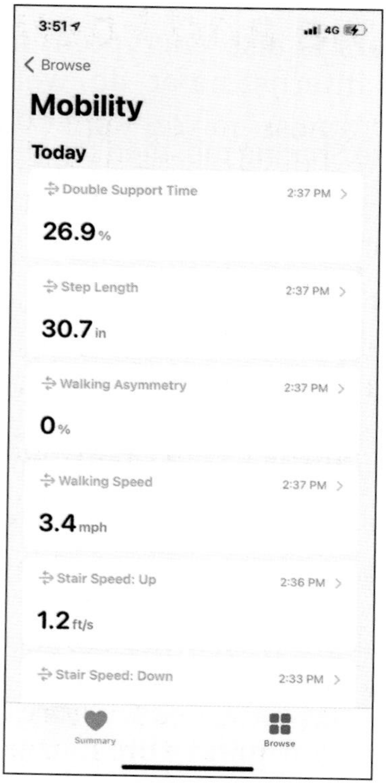

FIGURE 10-8

CONTRIBUTE TO HEALTH RESEARCH WITH YOUR APPLE WATCH

Apple is constantly working with researchers such as the American Heart Association, Harvard, Stanford, and the National Institute of Environmental Health Sciences to discover additional ways that technology can benefit humanity's health. You're invited to participate in cutting-edge research studies that use data from your Apple Watch and iPhone to provide crucial information to researchers. You can download and install on your Apple devices Apple's Research app to participate in studies such as the Apple Heart & Movement Study and the Apple Hearing Study. And yes, the data you provide is safe and private. To learn more, and to download the Research app, visit `www.apple.com/ios/research-app/`.

Just Breathe and Focus

The Mindfulness app on Apple Watch took me by surprise; it's such a simple idea but can provide major benefits, such as increasing focus while helping you take a moment to relax. In today's increasingly topsy-turvy environment, taking a few minutes to be mindful can only benefit us.

The app has two modes to help you center and practice better breathing: reflect and breathe.

Reflect prompts you with an idea or topic to think about, then provides a gorgeous animation to soothe you while you ruminate. This is a perfect way to get yourself back on track when you begin to feel tired or are losing interest in the tasks at hand or simply want to take a moment to reflect on something good, true, and beautiful.

Breathing is also a good thing — as someone who's lived with asthma since the age of 5, I can vigorously attest to this. However, breathing better is a great thing, especially when it's done in a mindful way. Just follow the on-screen prompts and breathe along with the beautiful animation and haptics (if enabled) for the duration of your breathing session.

Let's check out the Mindfulness app:

1. Open the Mindfulness app on your Apple Watch.

2. Rotate the digital crown to switch between Reflect or Breathe, as I'm doing in **Figure 10-9**.

3. Tap the more icon (three dots) to adjust the length of your reflect or breathe session, tap Duration, and then tap the length of time you'd like for the session (between 1 and 5 minutes). Tap Duration in the upper-left corner, and then tap Reflect or Breathe in the upper-left corner to return to the main Mindfulness screen.

4. Tap the Reflect or Breathe button to begin the session.

5. Follow the on-screen instructions to progress through and complete your session.

FIGURE 10-9

The Mindfulness app offers several settings so you can customize your experience. Open the Settings app on your Apple Watch and tap Breathe to view the following options:

» **Reminders:** Decide how many times you'd like your Apple Watch to remind you to participate in reflect or breathe sessions.

» **Weekly Summary:** Toggle on (green) to see an overview of the past week's sessions every Monday.

» **Mute for today:** Toggle on (green) if you want your Apple Watch to skip reminders for the day.

» **Breath Rate:** Select how many breaths per minute you'd like breathe mode to pace you with (7 is default).

» **Haptics:** Determine how haptic feedback is provided during your sessions. I find haptics helpful, but some folks consider it a distraction, so Apple is kind enough to let us decide how or if we want to use them.

» **Add New Meditations to Watch:** Toggle on (green) if you'd like new meditations to be downloaded to your Apple Watch when it's connected to power and near your iPhone. When you complete a meditation, it's deleted from your Watch automatically.

TIP

The breathe mode of the Mindfulness app even comes with its own watch face! The breathe watch face keeps the breathe mode of the Mindfulness app handy, allowing you to begin a session just by tapping the screen. If you want to simply focus on your breathing without beginning a new timed session, the animation on the watch face is running at the breathe rate set in Settings ⇨ Mindfulness.

Heartbeat City

When it comes to health tracking, the Apple Watch's original claim to fame was and remains its capability to monitor your heart rate. Whether working out, walking the dog, or sleeping, if you're wearing your Apple Watch, it will keep tabs on how fast your ticker's racing.

TECHNICAL
STUFF

How does Apple Watch monitor your heart rate? Well, that's the kind of magic of which we dare not speak in these parts . . . plus, the main word behind it all, *photoplethysmography*, is tough to pronounce. However, Apple is more than happy to share that info with you. Check out `https://support.apple.com/en-us/HT204666` to learn how flashing LED lights and electrical signals are part of the action.

The Heart Rate app is good as what it does, but your results can be affected by many factors. First, make sure that your Apple Watch is fitting your wrist properly — too loose or too tight and the readings won't be as accurate.

TIP

Please check out the following article at Apple's Support site for information on other factors that may affect your Apple Watch's accuracy with heart rate monitoring: `https://support.apple.com/en-us/HT207941#heartrate`.

Check your heart rate

Ready to check that heart rate?

1. Open the Heart Rate app on your Apple Watch.

   The app immediately begins measuring your heart rate, as shown in **Figure 10-10**. When it has finished measuring, your current heart rate is displayed.

2. Rotate the digital crown and you can see your heart rate while resting, walking, and while using the Mindfulness app.

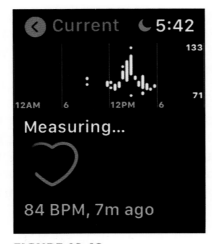

FIGURE 10-10

Receive heart health notifications

Not only will Apple Watch keep track of how your heart is working, it can also send you notifications when things may not be working as expected. The Heart Rate app can let you know when your heart rate may be too high or low, and also when you may be experiencing irregular rhythms.

To enable notifications for high and low heart rates:

1. Open the Watch app on your iPhone.

2. Tap the My Watch tab at the bottom of the screen.

3. Tap Heart in the menu.

4. Tap High Heart Rate, select a rate from the list, and then tap Heart in the upper-left corner.

5. Tap Low Heart Rate, select a rate from the list, and then tap Heart again in the upper-left corner.

TIP

Check with your physician if you're unsure what rates to set for your high and low heart rates.

To enable notifications for irregular heart rhythms:

1. Open the Health app on your iPhone.

2. Tap the Browse tab at the bottom of the screen.

3. Tap Heart in the menu, and then tap Irregular Rhythm Notifications.

   It's a good idea to read the information on the next screen before proceeding.

4. Tap the blue Set Up button.

5. Tap the Continue button at the bottom of the Apple Watch Can Look for Atrial Fibrillation screen.

6. Enter your date of birth, answer Yes or No to the question regarding whether you've been diagnosed with A-fib by a doctor, tap Continue, and then tap Continue again on the How It Works screens (after you've read the info, of course).

7. Read each item on the next screen, which shares Four Things You Should Know. Tap Next to proceed, and then tap Continue.

8. Finally, tap Turn on Notifications at the bottom of the screen to enable this great feature.

Now that you've enabled the irregular rhythm notifications, you can turn them on or off using the Watch app on your iPhone. Just open the app, make sure you're in the My Watch tab, tap Heart, and then toggle the Irregular Rhythm switch (shown in **Figure 10-11**).

FIGURE 10-11

TIP

I implore you to read Apple's Support article discussing heart health notifications: https://support.apple.com/en-us/HT208931. You'll find some important information about how your Apple Watch can help track A-fib. The article also explains a few things Apple Watch can't do, one of which is that it can't detect heart attacks (which is good to know, if you ask me).

Measure Blood Oxygen Levels

Apple Watch Series 6 and later can measure the oxygen level of your blood. The Blood Oxygen app uses the sensors on the back of your Watch to see just how bright or dark your blood is; the brighter the better when it comes to blood oxygen saturation.

REMEMBER

You must have an Apple Watch Series 6 or later and at least iOS 15 on your iPhone to use the Blood Oxygen app. Anything other than that simply won't work.

To get started with the Blood Oxygen app, you must first enable the feature in your iPhone's Health app:

1. Open the Health app on your iPhone.

2. If you see a prompt to enable Blood Oxygen, just follow the directions. Then continue after the steps.

3. Tap the Browse tab at the bottom of the screen.

4. Tap Respiratory ⇨ Blood Oxygen ⇨ Set up Blood Oxygen.

5. Follow the on-screen instructions to progress through the steps to enable the feature.

Now, let's measure some blood oxygen levels:

TIP

It's crucial that you remain as still as possible while the Blood Oxygen app works its magic. A twitch of the wrist or a tap on your Apple Watch's screen can throw the whole thing out of kilter — and you'll need to start again. I speak from experience.

1. Open the Blood Oxygen app on your Apple Watch.

2. Rest your arm on something and remember to keep as still as possible.

3. Tap the Start button and you'll see a countdown, as shown in **Figure 10-12**.

   The measurement takes 15 seconds, so stay as still as possible for that long. The measurement then appears on your Apple Watch screen.

4. You can view this and past measurements in your iPhone's Health app by going to Browse ⇨ Respiratory ⇨ Blood Oxygen.

FIGURE 10-12

Electrocardio-what?

Yet another arrow in Apple's healthcare quiver is the ECG app. Yep, your Apple Watch (Series 4, Series 5, Series 6, and Series 7 models, that is) can record a single-lead electrocardiogram, which is a measurement of your heart's electrical activity. Mind you, it's no replacement for the 12-lead ECG your doctor would give you, but the Apple Watch version is incredibly accurate. What's more, after you've recorded an ECG, you can easily share it with your physician if need be.

Like many other features in this chapter, you must first enable ECG using the Health app on your iPhone:

1. Open the Health app on your iPhone.

2. If you see a prompt to enable ECG, just follow the on-screen instructions. Then continue after the steps.

3. Tap the Browse tab at the bottom of the screen.

4. Tap Heart ⇨ Electrocardiograms (ECG) ⇨ Set up ECG App to begin.

5. Follow the on-screen instructions until you've enabled the feature.

Next, let's get busy recording your heart's electrical impulses:

TIP

While recording the ECG, don't sit close to an electronic device plugged into an electrical outlet, because electrical interference may play havoc with the results. I don't mean that you should be outside in the middle of a field, but you should perhaps be at least a few feet away from said electronics.

1. Open the ECG app on your Apple Watch.

2. Rest both of your arms on something and try to remain as relaxed and still as possible.

3. Using the hand opposite your Apple Watch, hold your finger on the digital crown (don't press the digital crown, just hold your finger against it) and you'll see a countdown, like the one in **Figure 10-13**.

   The recording takes 30 seconds, so stay as still as possible for that long. After the recording is complete, the results appear on your Apple Watch screen. Sinus Rhythm is good, all others are suspect and should be discussed with your physician.

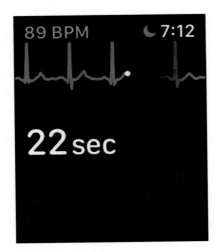

FIGURE 10-13

REMEMBER

No matter what results you received in the ECG app, if you're not feeling well, it's best to seek the advice of a medical professional ASAP.

You can view and share your ECG recordings using your iPhone's Health app:

1. Open the Health app on your iPhone.

2. Tap the Browse tab at the bottom of the screen.

3. Tap Heart ⇨ Electrocardiograms (ECG) and then tap a recording to see the full results, as shown in **Figure 10-14**.

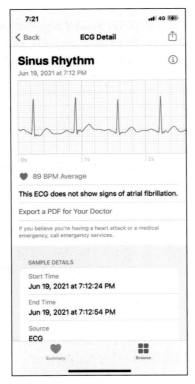

FIGURE 10-14

4. Share the recording with a medical expert:

   a. Export the PDF: Tap the blue Export a PDF for Your Doctor option (refer to Figure 10-14).

   b. Send the PDF to your physician: Tap the share icon in the upper-right corner (a box with an arrow pointing up), and then use one of the tools available, such as Mail or Messages.

TIP

Apple's Support site has an excellent in-depth article that delves into the nitty-gritty of recording ECG's with your Apple Watch: https://support.apple.com/en-us/HT208955. I suggest checking it out so you can get a fuller picture of what the results mean and how the technology behind the ECG app works.

Measure Noise Levels

Sometimes life is loud. Football games, concerts, crying babies, crying sports fans . . . occasionally the space around us can get downright eardrum-piercing. That's fine and dandy from time to time, but when it's sustained it can be harm your hearing. Your Apple Watch has a neat little app that uses the microphone of your Apple Watch to measure the ambient sound levels around you. The app alerts you when things may be getting a bit out of hand volume-wise.

TECHNICAL STUFF

No worries: The Noise app doesn't record anything. It only measures sound levels. Your privacy is intact.

Measuring noise levels is as simple as it gets. Open the Noise app on your Apple Watch. If this is the first time you've opened it, tap the Enable button to get going. The app begins measuring the surrounding noise levels. A report appears based on how high the decibel levels are. If the sound levels are dangerous for a sustained time, a warning will appear, as shown in **Figure 10-15**.

The Noise app can alert you when the sound reaches or exceeds a specified decibel level for a sustained period of 3 minutes:

1. Open the Settings app on your Apple Watch.

2. Tap Noise, and then tap Noise Notifications.

3. Tap a decibel level and duration to set them as the default for notifications.

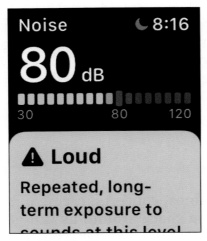

FIGURE 10-15

Catch Some Z's

Your Apple Watch isn't just for tracking health during waking hours; the Sleep app also helps you keep tabs on how you're resting during sleep. You can create schedules to help set the amount of time you want to sleep, set alarms to wake you, and view data collected while you sleep.

TIP

Your Apple Watch should have at least a 30 percent charge before you go to sleep. If not, it will prompt you to charge it before it can use the Sleep app.

To get started, you need to set up the Sleep app for the first time.

1. Open the Sleep app on your Apple Watch and tap Next.

2. Set the number of hours for your sleep goal by tapping the – and + buttons (see **Figure 10-16**), and then tap Next. (You may need to scroll down to see it.)

3. Create a schedule on the Set Your First Schedule screen:

   - Tap Active On and then select the days of the week the schedule applies to. Then tap Done.

- Tap the time in the Wake Up section to schedule your wake time. Tap inside the hour and minute windows, rotate the digital crown to select a time, tap AM or PM, and then tap Set, shown in **Figure 10-17**.

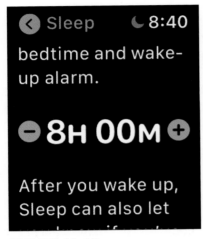

FIGURE 10-16

FIGURE 10-17

- Toggle on (green) the Alarm switch to enable an alarm sound and haptics.

- Tap the time in the Bedtime section to schedule your bedtime. Tap inside the hour and minute windows, rotate the digital crown to select a time, tap AM or PM, and then tap Set.

4. Tap Next to view your schedules.

If you'd like to create more than one schedule (for weekends, let's say), just tap the Add Another Schedule button and follow Step 3 again.

5. Tap Next, then either tap Enable to turn on Time Asleep Tracking or tap Skip to move to the next step.

Time Asleep Tracking detects your motion while your asleep. If you skip this step, you can enable it later in Settings ⇨ Sleep on your Apple Watch.

6. To turn on sleep mode for both your Apple Watch and your iPhone, tap Enable.

   Sleep mode helps prevent distractions during your sleep schedules. If you'd rather set this up for each device independently or skip it, just tap Skip. You can then enable sleep mode for either device in its respective Control Center.

7. Tap Next on the Charge Reminders screen, and then tap Done on the Summary screen. Sleep is set up and ready to go.

TIP

You can set up a Wind Down time, which sets a time for sleep mode to activate before your actual bedtime. Open the Sleep app on your Apple Watch, tap Full Schedule, and then tap Wind Down to set a time interval.

You can always adjust these options and schedules at any time by using:

» The Sleep app on your Apple Watch.

» The Sleep section of the Settings app on your Apple Watch.

» The Watch app on your iPhone. Go to the My Watch tab and tap Sleep.

» The Health app on your iPhone. Go to the Browse tab and tap Sleep. Tap Full Schedule & Options to see and modify the schedule you've already created, as shown in **Figure 10-18**.

TIP

You can view your sleep history by opening the Sleep app and then scrolling down to see how you slept the previous night and your average over the last couple of weeks. You can also view the same history on your iPhone's Health app by going to the Browse tab and tapping Sleep.

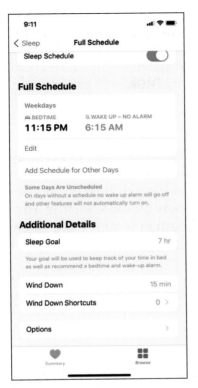

FIGURE 10-18

Fall Detection

When I describe the Fall Detection feature, many people are more pleasantly surprised by it than just about any other cool thing the Apple Watch can do. If you have a Series 4, Series 5, Series 6, SE, or newer, Apple Watch can detect when you experience a hard fall and — get this — even notify emergency services and your emergency contacts!

Here's how Fall Detection works. Let's say (heaven forbid) that you do take a hard fall. Apple Watch will detect it and tap your wrist with haptics, sound an alarm, and display a message that gives you the chance to call emergency services right away or let your Apple Watch know that everything's fine. However, if Apple Watch finds that you're not responsive and are immobile for roughly a minute, it will begin a 30-second countdown, the taps on your wrist will be persistent, and the alarm will become increasing louder. Once the 30 seconds is up, Apple Watch automatically alerts emergency services and your emergency contacts, even supplying them with your location if can determine it.

TIP

If you're someone who's active and regularly participates in high-impact exercise or play, you could accidentally trigger Fall Detection. If this happens, just tap the I'm OK button or the Close button on your Apple Watch display, or press the digital crown to dismiss the alert. Also, note that Apple Watch may not detect every single fall, although it's remarkably accurate.

For Fall Detection to work, it must be enabled on your Apple Watch:

1. Open the Settings app on your Apple Watch.

2. Tap SOS and then tap Fall Detection.

3. Tap to toggle on (green) the Fall Detection switch, as shown in **Figure 10-19**.

FIGURE 10-19

You can also enable Fall Detection on your iPhone:

1. Open the Watch app on your iPhone. Make sure you're in the My Watch tab; if not, tap My Watch at the bottom of the screen.

2. Tap Emergency SOS.

3. Tap to toggle on (green) the Fall Detection switch.

TIP

If you've entered your age when setting up your medical ID and health profile, and you're at least 55 years of age, Fall Detection will be enabled automatically.

» **Begin and end workouts**

» **Take a look at your workout statistics**

» **Get to know Apple Fitness+**

Chapter **11**

Get Moving!

ctivity is key to living a healthier and happier life. You don't need me to tell you this; it's something we all know deep in our bones. Even some tiny smidgen of activity, no matter what it is, is better than laying there like a limp noodle all day. Muscles contract and expand. The heart pumps and blood flows. Your skin glows and your mind focuses. Friends and family are suitably impressed.

But let's face facts — it's much easier to begin an activity and to stay active when we're held accountable. That's what coaches and mentors and teachers are for. We human beings need the urging and encouragement of another to help motivate us, especially when we're having "one of those days."

Apple Watch is no substitute for your yoga instructor, but it can be a good temporary helper, offering friendly reminders (and even rewards) to get moving and to reach your daily activity goals. In addition to tracking your activity, watchOS offers the Workout app, which keeps track of how you do in many different exercise types and scenarios.

In this chapter, you look at the ways your Apple Watch can help track your activities and get an introduction to Apple's great new service, Fitness+.

Let's do this! (That was my feeble attempt at sounding like a highly motivational personal trainer.)

WARNING

Before you begin any exercise routine, especially one that's outside your normal activities, check with your doctor. Your Apple Watch is a good motivator and a great way to keep records of your achievements, but it won't do much of either if you're not able to use it because you didn't make sure your body was up to the task beforehand.

The Lord of the (Activity) Rings

When the Apple Watch was first presented to the world, Apple introduced a nifty and catchy way to track your daily activity progress: closing your rings. You've probably seen the Apple Watch ads on TV with the red, green, and blue rings that rotate clockwise and close (see **Figure 11-1**); that's what that term *closing your rings* means.

REMEMBER

If your rings overlap, it's a good thing! That just means you've exceeded your activity goals for the day.

GET THE MOST ACCURATE READINGS POSSIBLE

The accuracy of how Apple Watch tracks your activity, especially the number of calories you burn through, is dependent on several factors. The most important of these is information you give it regarding your height, weight, age, and other factors. To find out more about how to get the most accurate readings possible with your Apple Watch, please check out Apple's Support site at https://support.apple.com/en-us/HT207941.

Image courtesy of Apple, Inc.

FIGURE 11-1

Move

The red ring is the move ring, and it shows how many active calories you've burned throughout the day. By keeping up with everything from your movements to heart rate, your Apple Watch can fairly accurately calculate how many calories you're burning. This ring closes once you've reached your move goal, which I describe later in the chapter.

Exercise

The green ring is the exercise ring. Apple Watch keeps track of how long you've been active at a pace "at or above a brisk walk." The exercise ring displays the number of minutes that you've maintained this activity. You'll close the ring after totaling at least 30 minutes of exercise activity.

Stand

The blue ring is the stand ring. Every time you stand and move for at least a minute an hour, you advance the stand ring; do this for any 12 hours in the day and you'll close this ring.

If you're someone who uses a wheelchair, make a simple change to your Apple Watch settings to convert the stand ring to the roll ring:

1. On your iPhone, open the Apple Watch app.

2. Tap the My Watch tab at the bottom of the screen.

3. Tap Health and then tap Health Details.

4. Tap Edit in the upper-right corner, and then tap Wheelchair.

5. Select Yes from the options that appear at the bottom of the screen.

6. Tap Done in the upper-right corner to lock in the setting.

Keep up with Daily Activity

watchOS includes an app to help you keep up with just how active you've been, and that app is where those glorious move, exercise, and stand rings reside. I'll bet you can't guess what the app's called.

Activity, of course! Apple does like to keep things simple, when possible.

Glance at the app

Open the Activity app by tapping its icon on your Apple Watch Home screen.

The Activity app begins the day with a clean slate, as shown **Figure 11-2**. However, as the day progresses and you can get out and about (I hope), the landscape of its contents begins to morph in front of your very eyes, as shown in **Figure 11-3**.

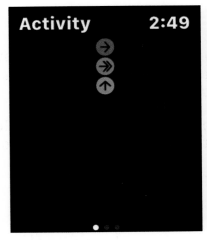

FIGURE 11-2 **FIGURE 11-3**

Dive a little more deeply

There's more to the Activity app than just pretty rings, though.

While viewing the rings, rotate the digital crown and you'll see a more detailed breakdown of your activity (see **Figure 11-4**) based on the three categories you're rapidly getting to know so well: move, exercise, and stand. You're shown

» An individual snapshot of how far each ring has moved

» A percentage of how much progress you've made toward your daily goal

» A fraction of how many calories (move), minutes (exercise), and hours (stand) you've achieved compared to the default goals

Rotate the digital crown a bit more and you're treated to a timeline of each activity, color-coded according to the each category, as shown in **Figure 11-5**.

As you continue to rotate the digital crown, you'll also find additional details of your activities, such as total steps, total distance, and flights climbed. (Yes, your Apple Watch knows when you're ascending and descending stairs!)

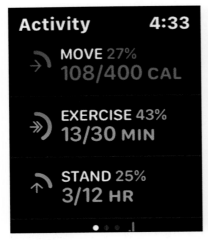

FIGURE 11-4

FIGURE 11-5

Finally, as you rotate the digital crown to the bottom of the page, you'll see two options: Weekly Summary and Change Goals.

View a summary of weekly activity

Tap Weekly Summary, and then rotate the digital crown (you could also swipe the display, if so inclined) to see

» A graph of how much total progress you made each day of this week

» Current totals for the following (some of which are shown in **Figure 11-6**): Calories, average calories, steps, distance, flights climbed, and active time

Tap Done to exit the weekly summary.

Change your weekly goals

Use the Change Goals button in the Activity app to manually adjust your weekly goals. This is a great idea as you progress in your fitness and want to set higher goals, or even if you've had a setback and need to lower your expectations a bit.

FIGURE 11-6

To adjust your activity goals:

1. Tap the Change Goals button in the Activity app.

2. Tap the + or – button in the Move Goal screen (see **Figure 11-7, left**) to adjust the number of calories you hope to burn each day for the week, and then tap the Next button.

3. Tap the + or – button on the Exercise Goal screen (see **Figure 11-7, middle**) to increase or decrease the number of minutes you plan to move at a brisk pace or better, respectively. Tap the Next button again.

4. Tap the + or – button (seeing a pattern emerge?) on the Stand Goal screen (see **Figure 11-9, right**) to change the number of hours each day that you plan to stand or move for at least a minute.

5. Tap the OK button to apply the changes to your goals.

TIP

Your Apple Watch prompts you to change your goals every Monday, and will offer suggestions based on your activity the previous week. You can accept the recommendations or set your own goals. And of course, you can manually adjust the goals at any time, as just outlined.

SHARE YOUR ACTIVITY GOALS

The Activity app and the Fitness app on your iPhone allow you to share your activity with friends and family. You can share with them, they can share with you, and if the notion strikes, you can even challenge each other through competitions. Once a competition starts, you'll both see each other's activity and also be alerted when you're in the lead or falling behind with your friendly competitor. To learn more about this great feature of Apple Watch, visit Apple's Support site at https://support.apple.com/en-us/HT207014.

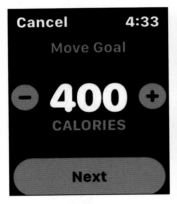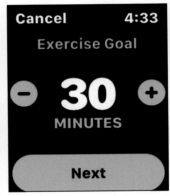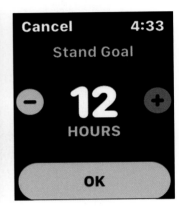

FIGURE 11-7

Explore a bit more with your iPhone's Fitness app

Your Apple Watch will sync the day's activity data with your iPhone when they're in proximity. You can then use your iPhone's Fitness app to see the activity data that's been stored there:

1. Tap to open the Fitness app on your iPhone.

2. Tap the Summary button at the bottom of the screen.

3. Tap the Activity area, as shown in **Figure 11-8**.

    On the next screen, you'll see a breakdown of the day's activities, color-coded in red (move), green (exercise), and blue (stand).

4. At the top of the screen, tap the initial for the day of the week you'd like to view: M, T, W, T, F, S(aturday), and S(unday), as shown in **Figure 11-9**.

FIGURE 11-8

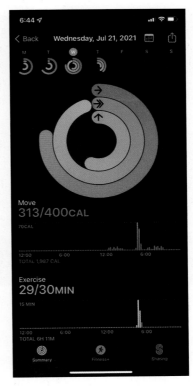

FIGURE 11-9

5. Tap the green calendar icon in the upper-right (refer to Figure 11-9) to see a calendar (as shown in **Figure 11-10**) displaying your daily and monthly progress.

6. Tap the Summary or Back button (the button that appears can vary depending on activities you perform) in the upper-left corner to return to the Fitness app's main screen.

7. Swipe down to the Trends area, shown in **Figure 11-11**, and tap a category to see more information.

   The Trends area compares the last 90 days of your activity with up to the previous 365 days.

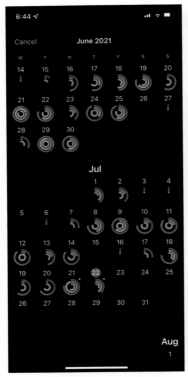

FIGURE 11-10

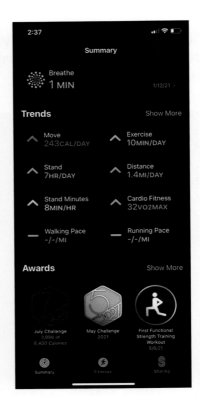

FIGURE 11-11

The category screen, shown in **Figure 11-12**, displays a graph comparing your progress in that category over the past year, as well as a graph showing a comparison of daily averages over that same span.

8. Tap the Summary button in the upper-left corner to return to the Fitness app's main screen.

"It's a major award!" — Mr. Parker, A Christmas Story

While Apple probably won't be sending you any leg lamps ("Fra-jee-lay. . . It must be Italian!"), they do look to bestow awards (major or otherwise) when you achieve your goals. Whenever you reach certain activity milestones, set personal records, or win friendly competitions, your Apple Watch will display a super–cool illustration as it reveals a badge for the award you've just earned.

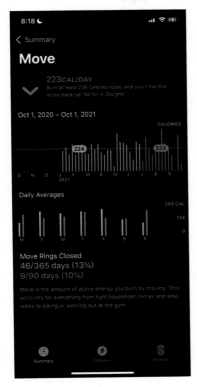

FIGURE 11-12

You can view your awards on your Apple Watch or on your iPhone.

To see awards on your Apple Watch:

1. Open the Activity app.
2. Swipe the display from right-to-left twice to view the Awards screen.
3. Rotate the digital crown to see a list of your awards.

4. Tap an award to see information about it, like I'm doing in **Figure 11-13**.

   You can also swipe the award to spin it and see information about when it was last achieved.

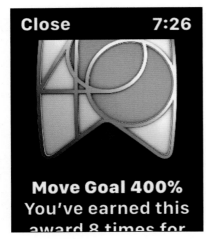

Close   **7:26**

Move Goal 400%
You've earned this
award 8 times for

FIGURE 11-13

To see awards on your iPhone:

1. Open the Fitness app.

2. Tap the Summary tab at the bottom of the screen.

3. Swipe down to the bottom of the page and tap Show More on the right side of the Awards section.

4. Tap an award to see information about it, as shown in **Figure 11-14**.

   You can also tap and hold down on the award, and then move your finger to spin it and see information about when you last achieved it.

FIGURE 11-14

Everything's Gonna Workout Just Fine

Working out is something all of us must do if we hope to stick around for as long as possible. Whether you're in the best shape of your life right now or the opposite is true, in most cases there's almost always something you can do to better your health. Working out doesn't have to involve hefting gargantuan weights; it can be something as simple as taking a stroll with your significant other.

The Workout app on your Apple Watch is a great way to keep tabs on your workouts, whether you've been working out for five decades or 5 minutes.

With Workout, you can choose from a big selection of workout types, and customize your workouts to meet certain goals, such as calories burned or distance. Just feel like winging it? You can simply start an open workout, which allows you to work out as long as you like without stopping after a specific goal is reached.

For an overview of your health and fitness endeavors, go to the Fitness app on your iPhone to view your workout data.

Start a workout

It's easy to get going with a workout:

1. Open the Workout app on your Apple Watch.

2. Rotate the digital crown to find the type of workout you'd like to participate in.

Don't see the workout you're looking for? Rotate the digital crown to the bottom of the list, tap the Add Workout button, and prepare to be stunned by the variety of workouts at your disposal!

3. Simply tap the workout type (Outdoor Walk is shown in **Figure 11-15**) to begin a countdown and start an open workout.

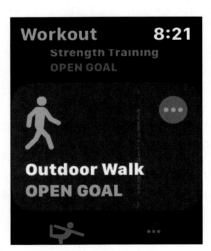

FIGURE 11-15

You can also begin a goal-specific workout:

1. Open the Workout app on your Apple Watch.

2. Rotate the digital crown to find the type of workout you'd like to participate in.

3. Tap the more icon (three dots) to the right of the workout type (refer to Figure 11-15).

4. Select the goals you'd like to use for this workout, as shown in **Figure 11-16**. Available options vary by workout type.

5. Tap the + or – button to adjust the goal settings, and then tap the Start button to begin, as shown in **Figure 11-17**.

FIGURE 11-16

FIGURE 11-17

Monitor a workout

You can easily check your progress during a workout. The Workout app should be open by default when you're in the middle of a workout. If you don't see it when you glance at your display or if you're using another app, do the following:

» If you're viewing your watch face, tap the icon for the Workout app at the top of your display.

>> Press the digital crown until you reach the Home screen, and then tap the Workout icon.

The progress of your workout is visible on the display. You can rotate the digital crown to highlight individual stats, such as elapsed time, active calories, total calories, and heartrate.

Pause or end a workout

Pause your workout if you come to an unexpected stopping point, or end your workout when you've reached your goals or are just plain tired.

To pause and end a workout:

1. Open the Workout app (if you're not already viewing your workout progress).

2. Swipe once from left-to-right to view your workout options (see **Figure 11-18**):

   - *Lock:* You can lock your display during a workout to prevent accidental bumps or taps (and to remove water or other fluids from the speaker), which could interrupt things. To unlock the display, rotate your digital crown until the Unlocked screen appears.

INTRODUCING FITNESS+

An Apple Fitness+ subscription offers workouts of all types led by some of the best personal trainers in the fitness world. Using the Apple Watch (it's required, actually), Fitness+ can also help you monitor your body's vital statistics while you work out, offering a new level of health tracking. Fitness+ also helps curate your workouts based on your previous exercise activities. You get a free 3-month subscription when you buy a new Apple Watch, or you can get a free 1-month trial by visiting Apple's Fitness+ website at www.apple.com/apple-fitness-plus/. If you like what you see, you can subscribe for $9.99/month. Better yet, you can bundle other great Apple services (like TV+ and News+) with Fitness+ and save over what you'd pay for them individually. Check out Apple's Apple One page at www.apple.com/apple-one/ for more information.

FIGURE 11-18

- *New:* Simply stop the current workout and start over.
- *Pause/Resume:* Tap to pause or resume your workout.
- *End:* Tap to end your workout.

When you end your workout, you're presented with a summary of your exertions. Swipe up and down on the display or rotate the digital crown to peruse the summary. After you're sufficiently impressed with your workout results, tap the Done button to dismiss the Summary screen.

Check your workout history

The Fitness app on your iPhone just loves to keep your activity and workout information so that you can check out your glorious accomplishments at a later date.

To view your workout history:

1. Tap to open the Fitness app on your iPhone.
2. Tap the Summary button at the bottom of the screen.

3. Tap the green Show More button to the right of the Workouts area, as shown in **Figure 11-19**.

4. Swipe up or down to see a list of your workout history by day and month.

5. Tap a workout to see details about it, as shown in **Figure 11-20**.

FIGURE 11-19

FIGURE 11-20

6. Tap Workouts in the upper-left to return to the Workouts screen.

7. To delete a workout from the list:

   a. Swipe the workout from right to left.

   b. Tap the red Delete button that appears to the right.

c. Determine whether to delete the workout and its data, or simply delete the workout, and then tap the corresponding option (see **Figure 11-21**).

FIGURE 11-21

4

A Media Extravaganza

IN THIS PART . . .

Taking and viewing pictures

Listening to audiobooks, music, and podcasts

Controlling Apple TV and music remotely

Chapter **12**

Shutterbugging

W hen I think of a watch, the idea of taking a photo with one isn't the first thing that pops into my mind. Back in the day, watches were for telling time, telephones were for making calls, and cameras were for taking photos. Then came the 2000s and things turned all topsy-turvy!

These days our phones include incredibly adept cameras. And now, thanks to Apple, we can pair our Watch with our phone to take pictures remotely. I always knew the future would be cool.

In this chapter, you learn how to use that beautiful Apple Watch on your wrist with that beautiful iPhone in your hand to take beautiful photos without having to hold said iPhone. Now you can take photos of yourself or someone else without having to strike the typical goofy-looking selfie pose.

You also look at using a photo album from the Photos app on your iPhone as an album to view on your Apple Watch. Then you find out how to take screenshots of what you see on your Apple Watch display — and more!

What's next, Apple? Phone calls from our Watch, too? Oh, wait . . . I think Chapter 6 may have mentioned something like that.

Take Pictures Remotely

Apple Watch comes with a nifty app called Camera Remote to facilitate the taking of remote photos using your iPhone. Your Apple Watch and iPhone connect to one another via Bluetooth, and the Camera Remote app on your Apple Watch directs the Camera app on your iPhone. You then use your Apple Watch's display to view what the iPhone camera is pointing at. Why would you want to do this? Because you can position your iPhone in places where you couldn't otherwise take a good picture, such as up on a shelf or on a counter, and then take the photo with your Watch.

REMEMBER

Be sure to keep your Apple Watch and iPhone within typical Bluetooth range of one another — about 33 feet — when using Camera Remote. Otherwise, the two will likely result in disconnect and you won't be able to remotely control your iPhone's camera.

TIP

It's a great idea to use a tripod (or even a tripod mount) designed to fit your iPhone when taking photos remotely. A tripod will securely hold your iPhone in place and reduce the chances of blurring in your photos. A tripod mount will allow you some versatility, letting you connect to multiple tripods or other threaded devices.

Capture the photos

Let's get started taking photos with your Apple Watch:

1. Open the Camera Remote app on your Apple Watch by tapping its icon (see **Figure 12-1**).

Your Apple Watch displays what your iPhone's camera is pointing at, as shown in **Figure 12-2**.

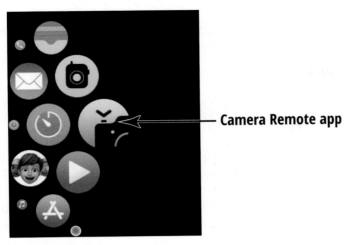 ———— **Camera Remote app**

FIGURE 12-1

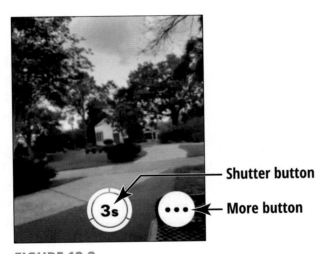

———— **Shutter button**

———— **More button**

FIGURE 12-2

2. Position your iPhone in the location where you want it to be when taking your photo.

WARNING

Make absolutely sure that your iPhone is secured in its location. If it falls, you could be looking at a very expensive repair as opposed to an expertly framed photo.

3. If you'd like to zoom in on your subject, simply rotate the digital crown on your Apple Watch.

4. When you're ready to take your photo, tap the shutter button (labeled in Figure 12-2).

   The picture is stored in the Photos app on your iPhone.

   By default, Camera Remote uses a 3-second delay to help you make sure you're in position for your photo, but this can be disabled. I show you how in a bit.

5. To review the photo on your Apple Watch, tap the thumbnail preview in the lower-left of the Apple Watch display (labeled in **Figure 12-3**).

6. When you have finished reviewing the image on your Apple Watch, tap Close in the upper left of the display.

Thumbnail image preview

FIGURE 12-3

Configure options

You can configure several options for the Camera Remote app. When you open the app, tap the more button (labeled in Figure 12-2) in the lower-right corner of the display. Rotate the digital crown or swipe up and down on the display to view all the possible options:

» *Timer:* As mentioned, Camera Remote includes a default 3-second delay. Simply tap to toggle the switch on (green) or off (see **Figure 12-4**), depending on your needs.

» *Camera:* Tap Front or Rear to choose which camera on your iPhone to use for the photo.

» *Flash:* Tap Auto, On, or Off (see **Figure 12-5**) to determine how to utilize the flash on your iPhone.

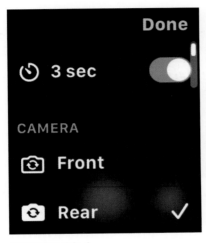

FIGURE 12-4

FIGURE 12-5

» *Live Photo:* Tap On or Off (see **Figure 12-6**) to enable or disable this feature, respectively.

» *HDR:* Tap On or Off to enable or disable High Dynamic Range (check **Figure 12-7**), which creates photos with greater luminosity (the strength of the light).

REMEMBER

HDR photos can be much larger in terms of data than standard photos, so they'll take up more of your iPhone's storage space.

Review photos you take with Camera Remote

As mentioned, you can review the photos you take using Camera Remote right on your Apple Watch display. This makes it a heck of a lot easier to see whether someone is rolling their eyes or you've cut off the top of their noggin in the shot without having to run back and forth to your iPhone.

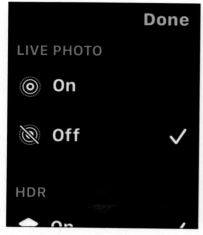

FIGURE 12-6

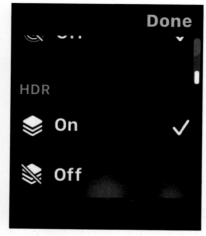

FIGURE 12-7

To review a shot in the Camera Remote app, simply tap the thumbnail image in the lower left of the Apple Watch display (refer to Figure 12-3). While viewing the image, you have a few more options:

» To view other images you've taken during this session, swipe right or left.

» Zoom in or out by rotating the digital crown.

» To pan (or move) around an image, zoom in and then drag the image around the Apple Watch display.

» If the photo doesn't fit the entire display but you'd like it to, simply double-tap the display. Double-tap again to return the photo to its original size.

» When you've finished reviewing the image(s) on your Apple Watch, tap Close in the upper left.

Take Screenshots of Your Apple Watch Display

A screenshot is basically a picture of what you see on your device's screen — in this case, your Apple Watch. Why might you want to take a screenshot? They can be extremely helpful when you have problems and need someone else to see what you're seeing. They're also good if you'd like to share or keep something like a fitness achievement or an alert that pops up on your display.

Screenshots can be especially helpful if you're writing a book about Apple Watch and want to show your readers what they should be seeing on their Apple Watch as they read along. Wink, wink.

WARNING
Be careful of the content in your screenshot if you plan on sharing it. In most cases, it's best to not share personal information, such as phone numbers, email addresses, and financial account numbers.

Taking a screenshot on your Apple Watch is one of the easiest things I'll show you how to do in this book: Simply press the digital crown and the side button at the same time, and the screen will flash (along with a shutter sound) to indicate you're successful. Done.

If you try to take a screenshot and it doesn't work, check to see if the feature is enabled in the Apple Watch Settings app. Go to Settings ⇨ General ⇨ Screenshots and toggle the Enable Screenshots switch on (green) if necessary (see **Figure 12-8**).

To view or share your screenshots, just use the Photos app on your iPhone, which is where all screenshots are automatically stored.

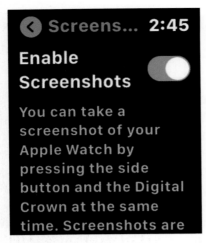

FIGURE 12-8

Work with Photos on Your Apple Watch

The Photos app on your Apple Watch works with the Photos app on your iPhone to allow you to sync and view photos between the two.

You can select which photo album in your iPhone's Photos app you'd like to use for your Apple Watch, and also determine how many photos to store on your Apple Watch, should you choose to do so.

Select a photo album and adjust storage

To select a photo album on your iPhone to sync with Apple Watch, and to determine how many photos to sync, do the following:

1. Open the Watch app on your iPhone.
2. Tap the My Watch tab at the bottom of the screen (see **Figure 12-9**).
3. Find and tap Photos (refer to Figure 12-9).

   The Photos screen appears, as shown in **Figure 12-10**.
4. Toggle the Photo Syncing switch on (green) or off.

Photos

My Watch tab

FIGURE 12-9

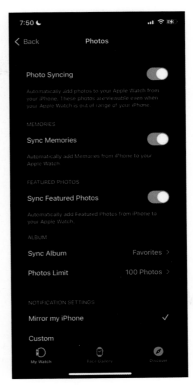

FIGURE 12-10

This option allows your iPhone to store photos from the Sync Album (see the next step) on your Apple Watch so you can view them even when your iPhone is out of range.

5. Tap the Sync Album option and select a photo album from your iPhone's Photos app to sync with your Apple Watch.

   The default album is Favorites, which consists of photos you've marked as favorites in your iPhone's Photos app.

6. Tap the Photos button in the upper left to return to the Photos screen.

7. To adjust how many photos are stored on your Apple Watch, tap Photos Limit.

8. Tap one of the options available (see **Figure 12-11**), and then tap the Photos button in the upper left to return to the Photos screen.

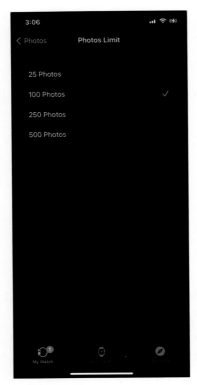

FIGURE 12-11

REMEMBER

If you need to change the number of photos stored on your Apple Watch at some point, don't worry if you change from a higher amount to a lower amount. Although photos may be removed from your Apple Watch, they will remain in the Photos app on your iPhone.

View photos

Your trusty Photos app on your iPhone has put in quite a bit of work in this chapter, but you're about to put its Apple Watch counterpart to work, too.

To view photos on your Apple Watch:

1. Open the Photos app on your Apple Watch by tapping its icon on the Home screen (see **Figure 12-12**).

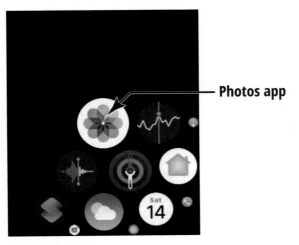

— **Photos app**

FIGURE 12-12

You see the photos that are part of the Sync Album you selected in the previous task, as shown in **Figure 12-13**.

2. If the photos are too small to make out clearly, rotate the digital crown to zoom in, as shown in **Figure 12-14**.

FIGURE 12-13

FIGURE 12-14

3. To view a photo, tap it.

4. While viewing a photo, try out the following:

   - Zoom in or out by rotating the digital crown.

   - Pan around the image by zooming and then dragging the image around the Apple Watch display.

   - Swipe right or left to view the previous or next photo, respectively.

5. To return to the album, rotate the digital crown until it zooms all the way out of the photo you're currently viewing.

TIP

To view a Live Photo on your Apple Watch, touch and hold down on the Apple Watch display while viewing the photo. For more information on Live Photos, visit `https://support.apple.com/en-us/HT207310`.

IN THIS CHAPTER

» **Explore and buy audiobooks**

» **Add audiobooks to your Apple Watch**

» **Play audiobooks with your Apple Watch**

Chapter **13**

~~Reading~~ Listening to Books

Apple has packed a lot into the tiny wonder we all know as Apple Watch. From news to weather to music to podcasts to photos to health apps and more, there's lots in there. And you know what else it needs? Books! Yes, books, but the audio type, of course.

Apple's been in the books business for years. The Apple Book Store serves a veritable plethora of digital and audio versions of books. Of course, reading a digital book on the tiny display of your Apple Watch would be more akin to a form of cruelty and punishment than enjoyment, so audiobooks is the way to go with your Apple Watch.

In this chapter, you discover how to find audiobooks for your Apple Watch, add them to your Apple Watch, and finally, listen to them using your Apple Watch and your favorite Bluetooth headphones.

Peruse the Apple Book Store

The first step to listening to audiobooks on your Apple Watch is to get some audiobooks to listen to.

The Apple Book Store has audiobooks in your favorite genres and at your favorite prices (even if that price is $0.00). The Audiobooks section of the Apple Book Store can be found in the Apple Books app on your iPhone. Once you've found and acquired a book, you'll then listen to it with your Apple Watch. Nice and tidy, that arrangement.

Shop for audiobooks

To shop using Apple Books:

1. Tap the Apple Books application icon on your iPhone to open the app. (It's on your first Home screen and looks like a white book against an orange background; it's also labeled Books.)

2. Tap the Audiobooks tab at the bottom of the screen.

3. In Audiobooks, shown in **Figure 13-1**, featured titles and suggestions (based on your past reading and listening habits and searches) are displayed. You can do any of the following to find an audiobook:

   • Tap the Search icon in the bottom right of the screen, tap in the Search field that appears, and then type a search word or phrase using the on-screen keyboard.

   • Swipe left or right to see and read articles and suggestions for the latest books in various categories, such as Featured Collection and Limited Time Offers.

   • Scroll down the Audiobooks main page to see links to categories of audiobooks such as Audiobooks for You and New & Trending, as shown in **Figure 13-2**. Swipe left and right on a category to view those selections, or tap the See All button under each to view even more.

   • Scroll down to Top Audiobooks to view bestselling audiobooks. Tap the See More Audiobooks button under Top Audiobooks to see the entire list.

   • Swipe further down the screen to find a list of Recent Bestsellers by Genre, as shown in **Figure 13-3**. Tap the See All button in each genre to see a complete list of bestsellers.

FIGURE 13-1

FIGURE 13-2

- Swipe down to the bottom of the Audiobooks page to find a list of genres, as shown in **Figure 13-4**. Tap All Genres to see all the Book Store has to offer.

- Tap a suggested selection or featured book to read more information about it.

You can also find audiobooks when browsing the Book Store tab of Apple Books. If an audiobook version of a title is available, you'll see a View the Audiobook link for it under the price, shown in **Figure 13-5**.

FIGURE 13-3

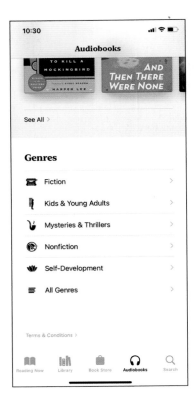

FIGURE 13-4

Try before you buy

Many books let you listen to a free preview before you buy. You get to hear a passage from the book to see whether it appeals to you, and it doesn't cost you a dime! To hear this preview:

1. While viewing the book's details, tap the Preview button under the audiobook price.

   A window appears near the bottom of the screen and the audio begins to play.

2. To view the progress of the audio sample, see the white line traversing the gray circle to the left of the book's title at the bottom of the screen, as shown in **Figure 13-6**.

3. To end the preview before it finishes, tap the square button in the gray circle.

FIGURE 13-5

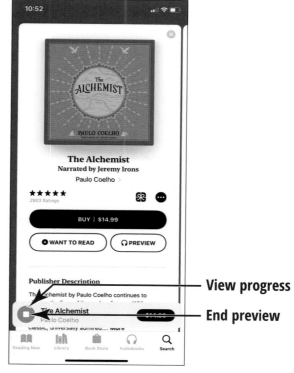

— View progress

— End preview

FIGURE 13-6

Buy Books

If you've set up a payment method in your Apple ID account (which you most likely have by this point), you can buy audiobooks in the Apple Book Store using the Apple Books app.

TIP

For more information on using payment methods with your Apple ID, please check out the article called "Change, add, or remove Apple ID payment methods" on Apple's Support site at https://support.apple.com/en-us/HT201266.

1. Open Apple Books, tap Audiobooks, and begin looking for a book.

2. When you find an audiobook that strikes your fancy, you can buy it by tapping it and then tapping the Buy button or the Get button (if it's free), as shown in **Figure 13-7**.

If you'd instead like to keep this audiobook in mind for a future purchase, tap the Want to Read button. Or tap Preview to hear a sample before you commit your hard-earned dollars to it (as discussed in the preceding section).

3. When the dialog appears at the bottom of the screen, purchase the book in one of the following ways:

   - *Tap Purchase, type your password in the Password field on the next screen, and then tap Sign In to buy the book.*

   - *Use Touch ID if you have it enabled for iTunes and App Store purchases. When you see the dialog telling you to Pay with Touch ID, place your finger on the Home button when prompted and follow the directions.*

   - *Use Face ID if you have it enabled for iTunes and App Store purchases. Follow the prompt on the right side of the screen and double-click the side button and glance at your iPhone to authenticate with Face ID.*

   The book begins downloading, and the cost, if any, is automatically charged to your account.

4. When the download finishes, tap OK in the Purchase Complete message. You can find your new purchase by tapping the Library button at the bottom of the screen, as shown in **Figure 13-8**.

TIP

Audiobooks that you've purchased in Apple Books can be accessed from any Apple iOS, iPadOS, or macOS device through iCloud.

TECHNICAL STUFF

Depending on where you are in the world, some items may not be available for purchase via Apple Media Services. If you're in the United States, you're most likely fine, but if not, please check Apple's Support site at `https://support.apple.com/en-us/HT204411` to find out more regarding product availability.

FIGURE 13-7

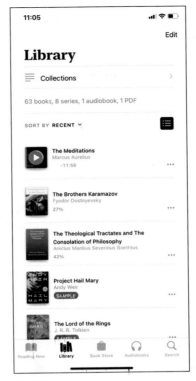

FIGURE 13-8

Sync Audiobooks to Your Apple Watch

Regardless of whether you purchased your audiobooks on an iPhone, an iPad, or a Mac, they'll be available to you for playback on your Apple Watch as long as you're using the same Apple ID.

By default, audiobooks from your Reading Now and Want to Read collections (found in the Apple Books app on your iPhone by tapping the Reading Now tab at the bottom of the screen, as shown in **Figure 13-9**) are automatically loaded on, or synced to, your Apple Watch. Up to 5 hours of the book you're currently listening to and up to 5 hours of the first book in your Want to Read collection are synced (assuming there's enough storage space available on your Apple Watch). Also, if you manually sync additional books, up to 5 hours of those are synced, as well (again, assuming space is available).

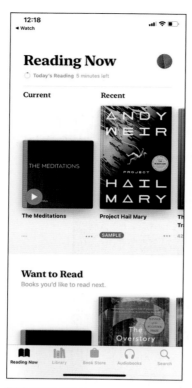

FIGURE 13-9

To add audiobooks to your Apple Watch:

1. Make sure your Apple Watch is connected to its charger before syncing books.

    If it's not, the sync will be queued but won't happen until the Apple Watch is back on its charger.

2. Open the Apple Watch app on your iPhone and tap My Watch at the bottom of the screen.

3. Find and tap Audiobooks.

4. The switches for Reading Now and Want to Read are both on (green) by default (see **Figure 13-10**), but you may turn them off (gray) if you want to prevent items in these categories from syncing to your Apple Watch.

5. To manually add additional audiobooks from your Apple Books library:

    a. *Tap the Add Audiobook button (refer to Figure 13-10).*

    b. *On the Select Audiobook screen, tap the title you want to add to your Apple Watch.* The new title appears in the From Library section, and the progress of its synchronization is indicated by an orange line moving clockwise around a circle (see **Figure 13-11**).

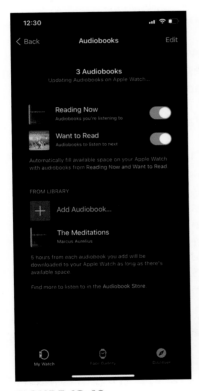

FIGURE 13-10

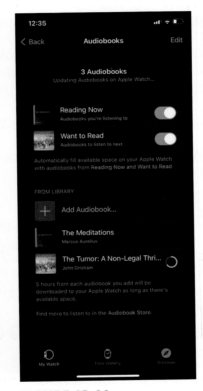

FIGURE 13-11

Listen to audiobooks without syncing

What if you don't want to add audiobooks to your Apple Watch? Maybe you don't have enough storage space on the watch or just don't want to use it for audiobooks. Whatever the reason, Apple's got you covered!

As long as your iPhone is near your Apple Watch, you can simply use the Audiobooks app on your Apple Watch to play audiobooks stored on your iPhone. I show you how in the upcoming section called "Play Your Audiobooks."

Delete audiobooks from Apple Watch

Maybe a book just didn't strike the right chord, or perhaps you've listened to it and want to free up space for another title, but at some point you'll want to delete audiobooks from your Apple Watch.

REMEMBER

Don't worry — deleting audiobooks from your Apple Watch does not delete them from other devices you may have downloaded them to, such as your iPhone or iPad.

To delete audiobooks from Apple Watch:

1. Open the Apple Watch app on your iPhone and tap My Watch at the bottom of the screen.

2. Find and tap Audiobooks.

3. In the From Library section, slide the title you want to delete from right to left to reveal the red Delete button, as shown in **Figure 13-12**.

4. Tap the Delete button to remove the title from your Apple Watch.

THIRD-PARTY AUDIOBOOK PROVIDERS

Although the Audiobooks app will play only titles you've purchased through the Apple Book Store on your iPhone, iPad, or Mac, other audiobook providers develop and supply apps so you can listen to their audiobooks on your Apple Watch. One such example is Audible. Other audiobook providers, such as Audiblebooks.com and Libby, offer iOS apps; you can control playback from your Apple Watch, but they don't provide a native Apple Watch app that allows you to sync content directly to it.

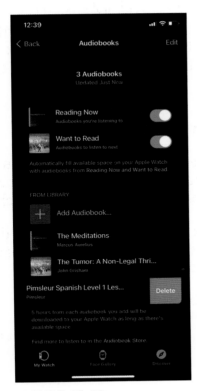

FIGURE 13-12

Play Your Audiobooks

The Audiobooks app, appropriately enough, is the app you'll frequent for your audiobook listening pleasure. Like its Apple Books counterpart on iOS, iPadOS, and macOS, its icon looks like an orange circle containing a white open book. Use it to play books that you've synced to your Apple Watch or that reside on your iPhone.

First things first! You need to pair Bluetooth headphones or earbuds to your Apple Watch before you can listen to audiobooks. When you first try to play an audiobook without a Bluetooth device connected, Apple Watch will prompt you to select one, as shown in **Figure 13-13**.

FIGURE 13-13

Refer to Chapter 6 for more info on connecting Bluetooth devices (Apple or otherwise) with your Apple Watch.

WARNING

The Audiobooks app will play only audiobooks you've acquired through the Apple Book Store.

To play audiobooks on your Apple Watch:

1. Press the digital crown until you access the Home screen.

2. Tap the Audiobooks app to open it on your Apple Watch.

   When you first open the app, you may see the cover image for the audiobook currently playing.

3. If necessary, rotate the digital crown or swipe up on the display to scroll through images of titles residing on your Apple Watch and to get to the main Audiobooks menu, as shown in **Figure 13-14**.

4. Choose the location the audiobook you want to listen to resides. Select from:

   - *On iPhone:* Audiobooks stored on your iPhone are listed.

   - *Library:* Audiobooks stored on your Apple Watch will be listed.

   - *My Family:* Audiobooks you've purchased for your family appear here.

FIGURE 13-14

5. Tap the title of an audiobook to begin playing it, as shown in **Figure 13-15**.

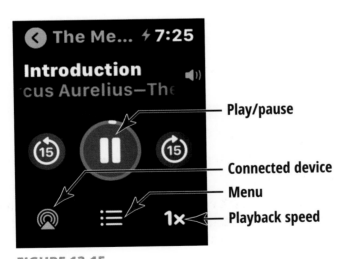

FIGURE 13-15

The options in the playback controls allow you to customize your listening experience:

» The progress in the track or chapter you're listening to is indicated by a white line that moves in a clockwise motion around the play/pause icon.

» To play, pause, and resume playback, tap the play/pause icon.

» To skip back 15 seconds in the audio, tap the button labeled 15 with the counterclockwise arrow (to the left of the play/pause icon). To skip ahead 15 seconds, tap the button labeled 15 with the clockwise arrow (to the right of the play/pause icon).

» To view and change which Bluetooth device you're listening with or to adjust any listed options (see **Figure 13-16)**, tap the connected device icon (triangle with radio waves) in the lower-left corner.

» To view the list of tracks or chapters the audiobook contains, tap the menu icon (three stacked lines) in the lower middle of the display. (See **Figure 13-17**.) Tap a track or chapter to play it if you want to skip around.

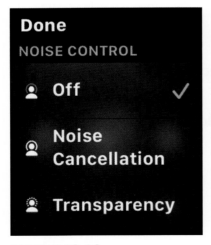

FIGURE 13-16

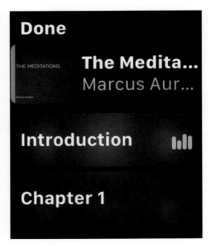

FIGURE 13-17

» To adjust how fast the audio plays, tap the playback speed icon (1x by default) in the lower-right corner of the display. When you tap the icon, it changes to reflect the speed you've selected.

» To adjust the volume, rotate the digital crown to adjust the volume. A green line appears to the right near the digital crown to indicate the volume level you're using, as shown in **Figure 13-18**.

» To return to the Audiobooks menu, tap the < in the upper left.

FIGURE 13-18

Chapter **14**

Groovin' on a Sunday (or Any Other) Afternoon

One of my favorite things to do with my Apple Watch is to play my favorite tunes and "shake my booty," as it were. I can't claim to be especially good at the aforementioned booty-shaking (I'm quite awful, if truth be told), but I can dance like nobody's watching if I pretend I'm Prince or Fred Astaire when the sounds move my soul.

Apologies to both of the superstars I mentioned if I've somehow tainted their legacies by association.

Using Bluetooth headphones or earbuds, your Apple Watch will let you listen to music, live radio, and podcasts — and in some cases, without your iPhone being in the near vicinity. Please commence to turning the pages in this chapter so that I may impart to you the magic of music (and other audio) played from the convenience of your wrist.

Music for the Masses

The Music app on your Apple Watch is your gateway to your tunes and other audio files. It's a miniature version of its iOS, iPadOS, and macOS iterations, and is just as easy (if not more so) to use.

In this section, I cover the devices that are required to listen to your music, where to get music, and how to start rockin' and rollin'.

How to listen

You'll need to pair Bluetooth headphones or earbuds to your Apple Watch before you can listen to audio. Without them, your Apple Watch will play the music on your iPhone, not through the watch itself.

Chapter 6 has much for more information on connecting Bluetooth devices with your Apple Watch. I can vouch that Apple's own AirPods Pro and AirPods are outstanding, but I'm also aware that some very fine third-party devices work well with Apple Watch and sound fantastic to boot.

Music sources

You can store music directly on your Apple Watch, play music stored on your iPhone, or stream music through your Apple Music subscription.

It's a bit outside the scope of this book to tell you how to get music (and other audio files) on your iPhone and to teach you all about the wonders of Apple Music. Other tomes (including two by yours truly: *iPhone For Seniors For Dummies* and *Apple One For Dummies*) cover both topics in detail. But I'm happy to share how you can store music directly on your Apple Watch so that you don't need to have your iPhone anywhere around to enjoy it.

You can store music on your Apple Watch in two ways: using the Apple Watch itself (you must have an Apple Music subscription) or using your iPhone.

TIP

You can't add individual songs to Apple Watch, regardless of the method you choose. You must add either playlists or albums.

Use Apple Watch to store music

You can add music to your Apple Watch from your Apple Watch if you have an Apple Music subscription. You can stream music (if you have an internet connection) or download music using this method.

REMEMBER

Your Apple Watch must be connected to its charger and Wi-Fi before it can download music with this method.

1. Open the Music app on your Apple Watch.

2. Tap Library, Listen Now, or Search, and then find the music you'd like to add.

3. Tap a playlist (shown in **Figure 14-1**) or album to view it.

4. Tap the more icon (three dots) on the right.

5. On the screen that appears, tap Add to Library to stream the playlist or album, or tap Download to download and store the music on your Apple Watch (see **Figure 14-2**).

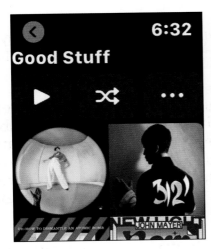

FIGURE 14-1

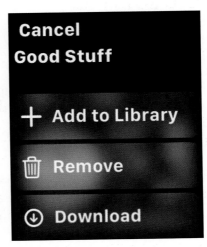

FIGURE 14-2

APPLE MUSIC AND MORE

Want to find out more about Apple Music, or sign up for a subscription? Go to www.apple.com/apple-music/ and learn all there is to know. Another option for signing up with Apple Music is through Apple One, Apple's service bundles that combine multiple Apple services at lower prices than you'd pay for each individually. To learn more, visit www.apple.com/apple-one/.

Use your iPhone to store music

You can also add music manually or automatically to your Apple Watch from the Apple Watch app on your iPhone.

REMEMBER

Your Apple Watch must be connected to its charger and near your iPhone before it can download music with this method. Your iPhone must also have Bluetooth enabled.

1. Open the Apple Watch app on your iPhone.

2. Tap My Watch at the bottom of the screen, and then tap Music.

3. To have content automatically added to your Apple Watch, toggle on (green) the option(s) in the Automatically Add section. (Recent Music is shown in **Figure 14-3**.)

4. To browse and select music from your iPhone, tap the Add Music button in the Playlists & Albums section (refer to Figure 14-3).

5. From the Library screen that appears (see **Figure 14-4**), find the playlists or albums you want to add to your Apple Watch.

6. Tap the tiny orange + button in the upper-right corner (see **Figure 14-5**) to add the playlist or album.

   The music will be added to your Apple Watch once it's connected to power and near your iPhone.

Remove music from Apple Watch

You may be running out of space for new music on your Apple Watch, or you may simply want to send a terrible song packing. Either way, at some point, you're probably going to need or want to remove music from your Apple Watch. You can do so directly from the Apple Watch (if you're an Apple Music subscriber) or from your iPhone.

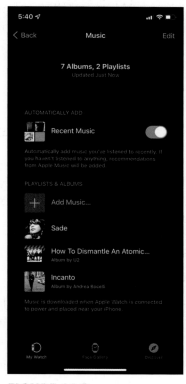

FIGURE 14-3

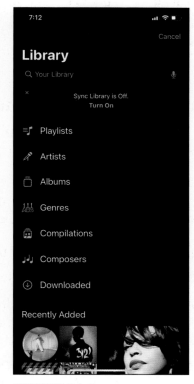

FIGURE 14-4

REMEMBER

Removing music from your Apple Watch does not necessarily delete it from other devices, such as your iPhone, iPad, or Mac. However, that is an option, as you'll see in the following steps, so be careful which option you choose.

Use your Apple Watch to remove music

To remove music from your Apple Watch by using your Apple Watch:

1. Open the Music app on your Apple Watch.

2. Tap Library and then tap Downloaded.

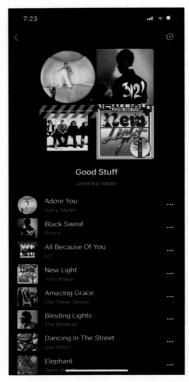

FIGURE 14-5

3. Tap a playlist or album to view it (see **Figure 14-6**).

4. Tap the more icon (three dots) on the right.

5. Tap Remove, and then tap either Remove Download or Delete from Library (see **Figure 14-7**).

WARNING

Remove Download simply removes the music from your Apple Watch — not from other devices. However, Delete from Library does remove the music from all devices using the same Apple ID. Be careful!

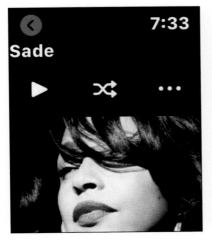

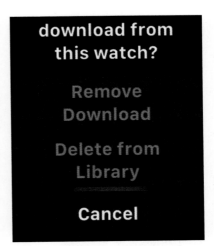

FIGURE 14-6

FIGURE 14-7

Use your iPhone to remove music

To remove music from your Apple Watch by using your iPhone:

1. Open the Apple Watch app on your iPhone.

2. Tap My Watch at the bottom of the screen, and then tap Music.

3. Toggle off (gray) the option(s) in the Automatically Add section (Recent Music is shown in **Figure 14-8**) to have this content removed from your Apple Watch.

   REMEMBER

   This content won't update again until Recent Music is turned back on.

4. In the Playlists & Albums section (refer to Figure 14-8), swipe the playlist or album you want to remove from right to left to reveal the red Delete button (shown in **Figure 14-9**), and then tap Delete.

REMEMBER

Music you remove from your Apple Watch using your iPhone will be removed only from the Apple Watch. It will remain on your iPhone.

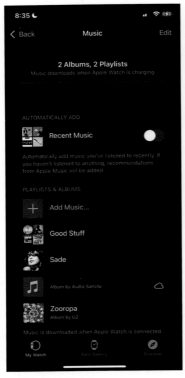

FIGURE 14-8

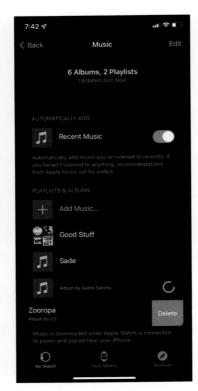

FIGURE 14-9

Play Your Music

The Music app is your go-to for playing music on your Apple Watch. Its icon is a red circle containing a white musical note.

REMEMBER

You'll need to pair Bluetooth headphones or earbuds to your Apple Watch before you can listen to music.

If you try to play music without a Bluetooth device connected, Apple Watch will prompt you to select one, as shown in **Figure 14-10**.

To play music on your Apple Watch:

1. Press the digital crown until you access the Home screen.

2. Tap to open the Music app on your Apple Watch.

FIGURE 14-10

3. When you first open the app, you may see album art. Rotate the digital crown or swipe up on the display to scroll through album art to get to the main Music menu, as shown in **Figure 14-11**.

4. Choose the location of the music you want to listen to. Select from:

   - *On iPhone:* You'll see music stored on your iPhone.
   - *Library:* You'll see music stored on your Apple Watch.

5. Find a playlist, album, or song you want to hear, and then tap it. If you tap a playlist or album, you'll need to tap the play icon (white arrow) on the left, as shown in **Figure 14-12**.

   After a song begins to play, the playback controls appear.

Here's how to control the playback of your audio files:

» To view the progress for the song you're listening, check out the white line that moves in a clockwise motion around the play/pause icon.

» To play, pause, and resume playback, tap the play/pause icon.

» To skip to the next song, tap the skip ahead icon (double white arrows, to the right of the play/pause icon).

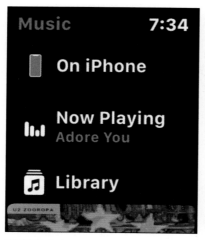

FIGURE 14-11

FIGURE 14-12

» To skip to the beginning of the song, tap the skip back icon (double white arrow to the left of the play/pause icon).

» To jump to the previous song, tap the skip back icon twice.

» To view and change which Bluetooth device you're listening with or adjust any options that may be listed (see **Figure 14-13**), tap the connected device icon (triangle with radio waves).

» To view the list of tracks in the Playing Next section, shown in **Figure 14-14**, tap the menu icon (three stacked lines). Then tap a track if you want to hear something different.

» To view the playback options at the top (refer to Figure 14-14), tap the menu icon (three stacked lines). Then tap the option(s) you want to use for the items in the Playing Next queue. From left to right, the options are

- *Shuffle:* Tap to play songs randomly.

- *Repeat:* Tap this once to repeat a playlist or an album, or tap twice to repeat a song.

- *Auto Play:* The Music app will automatically select songs to add to the end of the Playing Next queue based on the music you've already listened to.

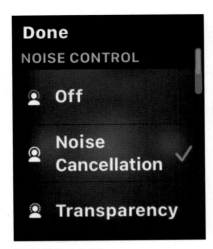

FIGURE 14-13

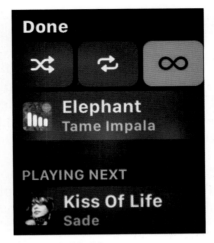

FIGURE 14-14

» To access additional options for the song, tap the more icon (three dots) in the lower-right corner of the display. Scroll through the options (some of which are shown in **Figure 14-15**) by swiping the display or rotating the Digital Crown, then tap one to utilize it.

» To adjust the volume, rotate the digital crown. A green line will appear to the right near the digital crown to indicate the volume level you're using, as shown in Figure **14-16**.

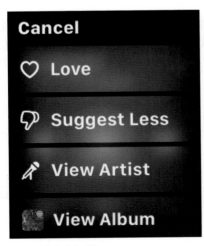

FIGURE 14-15

FIGURE 14-16

Raised on Radio

The more things change, the more they stay the same. While digital music has certainly taken its place in our culture, radio is still alive and well. So much so that Apple dedicated an entire app — the Radio app — to playing audio from that venerable technology on your Apple Watch.

The Radio app is at its best when you have an Apple Music subscription, but you can still use the app for broadcast radio stations without a subscription.

Listen to Apple Music radio

Apple does love their perks, and Apple Music subscribers have many to choose from. One such perk is access to specialized live Apple Music radio stations: Apple Music 1, Apple Music Hits, and Apple Music Country. Each station is designed to play the latest and greatest music in several categories. Another such perk of Apple Music is a collection of radio stations that are curated by musical experts and cater to very specific genres.

REMEMBER

Your Apple Watch need to be connected to Bluetooth headphones or earbuds to listen to stations with the Radio app.

To listen to Apple Music radio:

1. Press the digital crown until you access the Home screen.

2. Tap the Radio app icon (a red circle containing radio waves) to open the app on your Apple Watch.

3. Rotate the digital crown or swipe up on the display to scroll to the main Radio menu, as shown in **Figure 14-17**.

4. Tap the Stations button.

5. Tap one of the Apple Music stations (Apple Music 1, Apple Music Hits, or Apple Music Country) to take a listen.

You can also swipe the display or rotate the digital crown to find and play another genre-specific station. I'm giving Apple Music Country a spin in **Figure 14-18**; the playback controls appear after you tap a station.

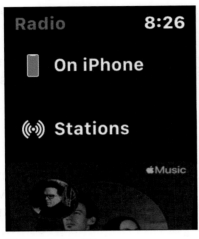

FIGURE 14-17

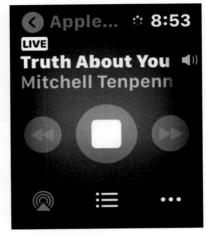

FIGURE 14-18

Use the playback controls to do the following:

>> Play, stop, and resume playback by tapping the play/stop icon.

>> View and change which Bluetooth device you're listening with, or adjust any options that may be listed by tapping the connected device icon (triangle with radio waves).

>> Access additional options for the song currently playing on the station by tapping the more icon (three dots). Scroll through the options (some of which are shown in **Figure 14-19**) by swiping the display or by rotating the digital crown, and then tap an option to use it.

>> Adjust the volume by rotating the digital crown. A green line will appear on the right near the digital crown to indicate the volume level you're using, as shown in **Figure 14-20**.

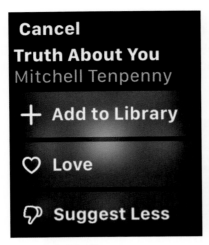

FIGURE 14-19

FIGURE 14-20

Listen to broadcast radio

You don't need an Apple Music subscription to listen to broadcast radio. However, there are so many stations out there that Apple can't list them all. You'll need Siri's help to find the station you want to listen to.

To listen to broadcast radio, simply ask Siri to play the station you'd like to listen to. Say something like "Hey, Siri, play 92.1 WZEW" or "play NPR."

After Siri finds the station, the playback controls open and the station begins to play automatically (see **Figure 14-21**).

In the playback controls, tap the play/stop icon to play, stop, and resume playback. Tap the connected device icon (triangle with radio waves) to view and change which Bluetooth device you're listening with, or to adjust any options that may be listed.

FIGURE 14-21

Podcast People

Podcasts (which are like radio shows, except you can listen to them any freaking time you want!) have become a cultural phenomenon all on their own. I could be off by a bit, but I believe that pretty much everyone and their pets have podcasts in which they discuss everything from politics to movies to plants to any and every other thing on the planet (and beyond).

Subscribe to podcasts

The Podcasts app on your iPhone is the tool you'll use to find and listen to podcasts on your iPhone or Apple Watch. To search Apple's massive library of podcasts and subscribe to them (which is insanely and awesomely free!), follow these steps:

1. Tap the Podcasts app on your iPhone to open it.

2. Discover podcasts in the following ways:

   - Tap Browse at the bottom of the screen. You'll find podcasts featured by Apple in categories such as Subscriber Favorites, as shown in **Figure 14-22**.

- Tap Browse at the bottom of the screen, and then swipe all the way down to Top Charts. Tap See All, and you'll see lists of the most popular podcasts. Tap All Categories in the upper-right corner to sift through the podcasts based on category (such as News, Religion & Spirituality, and Science), as shown in **Figure 14-23**.

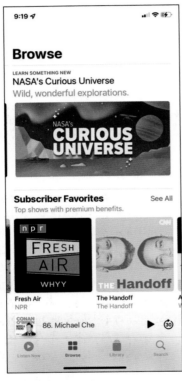

FIGURE 14-22

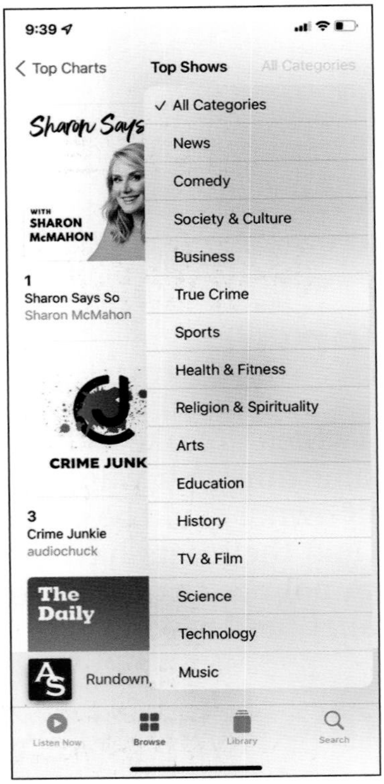

FIGURE 14-23

- Tap Search and then tap the Search field at the top of the screen. When the keyboard appears, type the name or subject of a podcast to see a list of results, which begin to appear as you type.

3. When you find a podcast that intrigues you, tap its name to see its information page, such as the one in **Figure 14-24**.

4. Tap the subscribe icon (+) in the upper-right corner.

   The podcast appears in the Library section of the app, and the newest episode is downloaded to your iPhone.

5. Tap Library in the toolbar at the bottom of the screen, tap the name of the podcast you subscribed to, and view its information screen.

6. Tap the more icon (three dots) in the upper-right corner and then tap Settings to see the settings for the podcast. From here (see **Figure 14-25**), you can customize how the podcast downloads and organizes episodes.

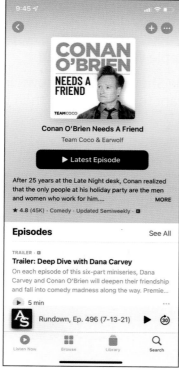

FIGURE 14-24

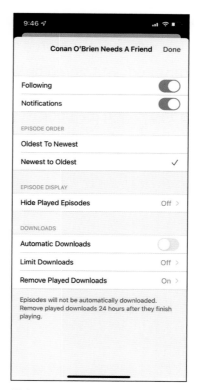

FIGURE 14-25

7. Tap Done in the upper-right corner when you're finished with the Settings options.

Play podcasts

Once you've gotten some podcasts on your iPhone, playing them with your Apple Watch is super–simple and works much like playing music or listening to radio stations.

To play podcasts on your Apple Watch:

1. Press the digital crown until you access the Home screen.

2. Tap to open the Podcasts app on your Apple Watch.

3. Rotate the digital crown or swipe up on the display to scroll to the main Podcasts menu, shown in **Figure 14-26**.

4. Choose the location of the podcast you want to listen to. Select from:
   - On iPhone: Podcasts stored on your iPhone are listed.
   - Library: Podcasts stored on your Apple Watch are listed.

5. Rotate the digital crown to scroll to the Shows section, shown in **Figure 14-27**. Find the podcast you want to listen to, and then tap it.

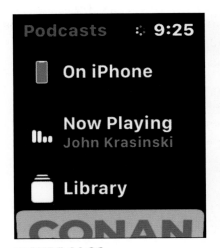

FIGURE 14-26

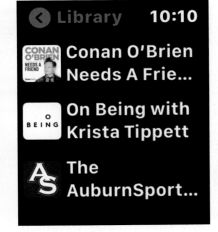

FIGURE 14-27

6. Tap the episode of the podcast you want to listen to.

   After the podcast begins to play, the playback controls appear.

The following options are available when playing podcasts:

» To view the progress for the episode you're listening to, see the white line that moves in a clockwise motion around the play/pause icon.

» To play, pause, and resume playback, tap the play/pause icon.

» Tap the button labeled 15 with the counterclockwise arrow (to the left of the play/pause icon) to skip back 15 seconds in the audio. Tap the button labeled 30 with the clockwise arrow (to the right of the play/pause icon) to skip ahead 30 seconds.

» Tap the connected device icon (triangle with radio waves) to view and change which Bluetooth device you're listening with, or to adjust any options that may be listed.

» Tap the menu icon (three stacked lines) to view the list of episodes for the current podcast in the Playing Next section, shown in **Figure 14-28**. Tap an episode to play it.

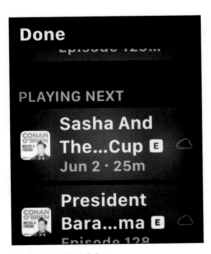

FIGURE 14-28

THIRD-PARTY MUSIC AND PODCAST OPTIONS

There are other music and podcast services on the globe (don't tell Apple I told you, please), so don't be shy if you prefer using something else. Examples are Spotify, Amazon Music, and Overcast. Search the App Store on your iPhone to find apps for these alternative services (and others), and chances are they'll have a companion Apple Watch version of their app.

» Tap the playback speed icon (1x by default) to adjust how fast the audio plays. When you tap the icon, it changes to reflect the speed you've selected.

» Rotate the digital crown to adjust the volume. A green line appears on the right near the digital crown to indicate the volume level you're using.

Chapter **15**

Getting Things Under Control

O h, how I pine for the good old days. You know, when the remote control in the home was you. Your mom or dad would say, "Change it to the *Andy Griffith Show* on channel 5," and you'd trudge from the couch to the television and flip the knob until the perpetually fuzzy channel 5 popped up on the screen.

Okay, after thinking about it, I don't pine for those days. Not at all. Nope.

Now, please don't get me wrong — I still like the *Andy Griffith Show*, but I let my Apple Watch do the heavy lifting of remotely controlling what we play on our Apple TV. Now I can enjoy the greatest show in the history of history (don't even argue with me) without wearing a trail on the carpet between the couch and the telly.

Life is good.

If you like, I'll be glad to show you how it's done. I'll also show you how to remotely control music on your computer, whether it's a Mac or a PC, using that trusty Apple Watch on your wrist. Just keep reading and you'll be living the good life, too.

Remotely Control Your Apple TV

The Remote app that comes with the latest version of watchOS on your Apple Watch allows you to use its display to remotely control your Apple TV. You can use your Watch's display to move among the apps and controls of your Apple TV in almost the same way you can use the Apple TV's own remote.

The Remote app, shown in **Figure 15-1**, can be found lounging around in the Apple Watch Home screen.

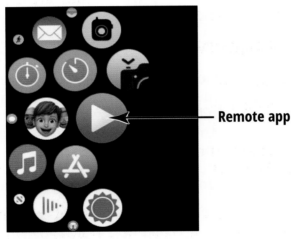

Remote app

FIGURE 15-1

And yep, you guessed it — for this section, you'll need an Apple TV to follow along. Or if you'd just like to see how it's done before springing for one of those delightful black boxes, please feel free to read on.

To learn more about Apple TV, visit www.apple.com/tv.

APPLE TV+ IS A DEFINITE PLUS!

You'll want to check out Apple TV+, which is a subscription-based content provider from Apple. It's quickly become one of my go-to services for entertainment, with such shows as *Ted Lasso* and *Foundation*, along with movies like *Greyhound*. Learn more about it by visiting `www.apple.com/apple-tv-plus`.

REMEMBER

Your Apple TV and Apple Watch must be connected to the same Wi-Fi network for remote control to work.

Pair your Watch with Apple TV

Your Apple Watch and Apple TV need to be introduced to one another (or paired, as we tech nerds call it) to work fruitfully together to enhance your remote control experience.

To pair Apple Watch and Apple TV:

1. Open the Remote app on Apple Watch.

2. If you see your Apple TV listed on the Apple Watch display, tap it and continue to Step 4.

3. If you don't see your Apple TV listed, tap the Add Device button, shown in **Figure 15-2**, then tap the Apple TV that's displayed (if one is found).

    If one isn't found, make sure your Apple TV is turned on and connected to the same Wi-Fi network as your Apple Watch.

4. On your Apple TV, open the Settings app (shown in **Figure 15-3**).

    The Settings app may be in a different location on your Apple TV Home screen, so just be on the lookout for the gear icon.

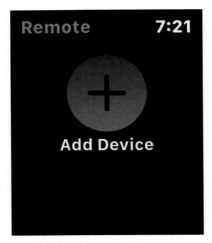

FIGURE 15-2

Settings app

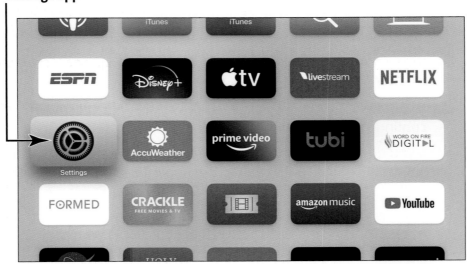

FIGURE 15-3

5. In the Settings app, select Remotes and Devices, shown in
Figure 15-4.

TIP

On older Apple TV models, you may need to choose different menu
commands. On an Apple TV 3, for example, choose General ⇨
Remotes to display the Remotes screen, and then select the Apple
Watch in the iOS Remotes list.

FIGURE 15-4

6. Select Remote Apps and Devices, like I'm doing in **Figure 15-5**.

7. Select Apple Watch.

8. When prompted by your Apple TV, enter the 4-digit passcode shown on your Apple Watch.

   An icon appears next to your Apple Watch when the two are paired.

FIGURE 15-5

Take control

Now that the two devices are on speaking terms, you can use your Apple Watch to control what goes on with your Apple TV.

REMEMBER Make sure your Apple TV is on and awake.

To control your Apple TV:

1. Open the Remote app on your Apple Watch.

2. Tap your Apple TV on the display, like the one in **Figure 15-6**.

   The Remote screen appears on your Apple Watch display, as shown in **Figure 15-7**.

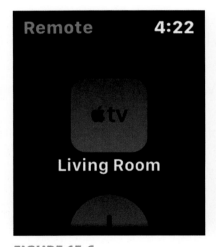

FIGURE 15-6

FIGURE 15-7

3. Control your Apple TV with the following actions:

   - Swipe up, down, left, and right to navigate the items on your Apple TV screen.

   - Tap to select an item, such as an app or menu option.

   - Tap the Menu button in the lower left of your Apple Watch display to go back one screen.

   - Tap the playback controls on your Apple Watch when viewing content.

Unpair your Watch and Apple TV

If you need to break up the beautiful friendship between your Apple Watch and Apple TV, you can do so easily:

1. Open the Settings app on your Apple TV.

2. Go to Remotes and Devices, and then Remote App and Devices.

3. Under Remote App, tap your Apple Watch, and then tap Unpair Device.

4. Open the Remote app on your Apple Watch.

5. Tap Remove (see **Figure 15-8**) when the *lost connection* message appears on the display.

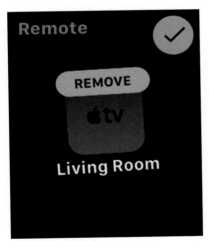

FIGURE 15-8

Control Music on Your Computer

The Remote app on your Apple Watch is kind of like a remote control Swiss Army knife — it's good at controlling more than just your Apple TV. It can also remotely control music stored on your Mac or Windows-based PC.

Your computer must be connected to the same Wi-Fi network as your Apple Watch, and the following conditions must be met:

» If you have a Mac running macOS 10.15 or newer, it uses the Music app to store and manage its music and other audio files, so it's ready to go.

» If you have Mac running macOS 10.14 and older, or a Windows-based PC's, it must use iTunes to store and manage its audio and music files.

TIP

If you don't have iTunes installed, go to `www.apple.com/itunes` and download it from there.

To control music on your Mac or PC with Apple Watch:

1. Open the Remote app on your Apple Watch, and tap the Add Device button.

   A four-digit passcode appears, like the one in **Figure 15-9**. I come back to the passcode in a moment.

FIGURE 15-9

2. On your Mac or PC:

- If you're using a Mac running macOS 10.15 or newer, open the Music app, and then select your Apple Watch in the devices list, as shown in **Figure 15-10**.

- If you're using a Mac or PC using iTunes, click the Remote button in the upper-left corner of the iTunes window.

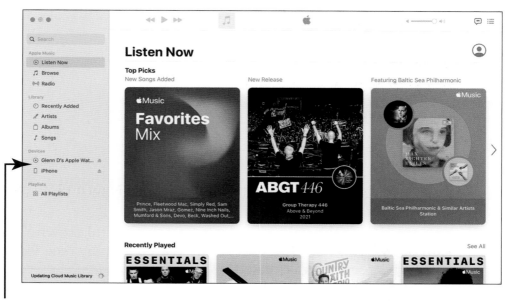

Devices list

FIGURE 15-10

3. When prompted in Music or iTunes, enter the four-digit passcode from your Apple Watch.

Music or iTunes confirms that you can control music remotely with your Apple Watch, as shown in **Figure 15-11**.

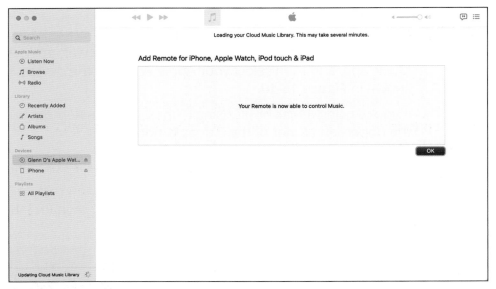

FIGURE 15-11

Use the playback controls on your Apple Watch (shown in **Figure 15-12**) to control the music playing on your computer.

You can adjust the volume by rotating the digital crown. The volume icon on the right side of your Apple Watch display (see **Figure 15-13**) will turn green as you make the adjustment.

FIGURE 15-12

FIGURE 15-13

Index

F

About the Author

Dwight Spivey has been a technical author and editor for over a decade, but he's been a bona fide technophile for more than three of them. He's the author of *iPhone For Seniors For Dummies, 2022 Edition* (Wiley), *iPad For Seniors For Dummies, 2022-2023 Edition* (Wiley), *Apple One For Dummies* (Wiley), *Idiot's Guide to Apple Watch* (Alpha), *Home Automation For Dummies* (Wiley), *How to Do Everything Mac* (McGraw-Hill), and many more books covering the tech gamut.

Dwight's technology experience is extensive, consisting of macOS, iOS, Android, Linux, and Windows operating systems in general, educational technology, learning management systems, desktop publishing software, laser printers and drivers, color and color management, and networking.

Dwight lives on the Gulf Coast of Alabama with his wife, Cindy, their four children, Victoria, Devyn, Emi, and Reid, and their pets Rocky, Penny, and Mirri.

Dedication

To Mom and Dad (Glinda and Glenn). I had you in mind during the writing of this book, especially since you both now have Apple Watches of your own. No son could be more grateful for your example of how to ride this big rock through the galaxy. Love you!

Author's Acknowledgments

Carole Jelen, my long-time agent; you make this whole thing work, and I appreciate you eternally!

Much gratitude goes to Steve Hayes, Susan Pink, and Guy Hart-Davis; each one of you are extraordinary at keeping me on track. And of course, the fantastic editors, designers, and other Wiley professionals who are truly critical to the completion of these books I'm so blessed to write. I want every individual involved at every level to know how grateful I am for their dedication, hard work, and patience in putting together this book.

Publisher's Acknowledgments

Executive Editor: Steve Hayes

Project Editor: Susan Pink

Production Editor: Mohammed Zafar Ali

Technical Reviewer: Guy Hart-Davis

Proofreader: Debbye Butler

Cover Image: ©wundervisuals/ Getty Images